Practical Guide Book Series™
Volume 3

ASME B31.3
Process Piping

Glynn E. Woods, P.E.
Roy B. Baguley, P.Eng.

Executive Editor
John E. Bringas, P.Eng.

Published By:

CASTI Publishing Inc.
14820 - 29 Street
Edmonton, Alberta, T5Y 2B1, Canada
Tel: (403) 478-1208 Fax: (403) 473-3359
E-mail: castipub@compusmart.ab.ca
Internet Web Site: http://www.casti-publishing.com

ISBN 0-9696428-4-9
Printed in Canada

THE PRACTICAL GUIDE BOOK SERIES™

Volume 1 - The Practical Guide to ASME Section II - Materials Index

Volume 2 - The Practical Guide to ASME Section IX - Welding Qualifications

Volume 3 - The Practical Guide to ASME B31.3 - Process Piping

Other Volumes in Preparation

First printing, May 1996
Second printing, July 1996
ISBN 0-9696428-4-9

FROM THE PUBLISHER

IMPORTANT NOTICE

OUR MISSION

Our mission at *CASTI* Publishing Inc. is to provide industry and educational institutions with practical technical books at low cost. To do so, each book must have a valuable topic, be current with today's technology, and be written in such a manner that the reader considers the book to be a reliable source of practical answers that can be used on a regular basis. *The Practical Guide Book Series*™ to the ASME Boiler & Pressure Vessel Code and Pressure Piping Codes has been designed to meet these criteria.

We would like to hear from you. Your comments and suggestions help us keep our commitment to the continuing quality of *The Practical Guide Book Series*™.

All correspondence should be sent to the author in care of:

CASTI Publishing Inc., 14820 - 29 Street,
Edmonton, Alberta, T5Y 2B1, Canada,
tel: (403) 478-1208, fax: (403) 473-3359
E-mail: castipub@compusmart.ab.ca
Internet Web Site: http://www.casti-publishing.com

ACKNOWLEDGMENTS

Glynn Woods would like to acknowledge the following people for their assistance in the preparation of his work: H.H. George, C. Becht IV, R.C. Hawthorne, L.A. Nuesslein, Jr., P. Ellenberger, E.C. Rodabaugh, F.R. Fairlamb, R.D. Hookway, D.R. Edwards, W.C. Koves, R.W. Haupt, K.L. Kluge, and J.S. Schmidt.

A special thank you is extended to J.P. Breen for assisting in reviewing the design portion of this book.

Grammatical editing and layout consulting was performed by Jade DeLang Hart. Graphics creation and photograph enhancements were performed by Denise Lamy.

These acknowledgments cannot, however, adequately express the publisher's appreciation and gratitude for their valued assistance, timely work, and advice.

DEDICATIONS

Dedicated to Sally, the love of my life.

Glynn E. Woods, P.E.
May 31, 1996

For Margaret, Andrew, and Robert, who tolerated the many evenings and weekends of work necessary for my contribution to this book.

Roy B. Baguley, P.Eng.
May 31, 1996

PREFACE

The ASME B31.3 Process Piping Code provides a minimum set of rules concerning design, materials, fabrication, testing, and examination practices used in the construction of process piping systems. However, B31.3 offers little explanation with respect to the basis or intent of the Code rules. Occasional insight can be gleaned from published interpretations of the Code, but these interpretations are answers to very specific questions asked by Code users. Any conclusions regarding the basis or intent of Code rules must be derived or inferred from the interpretations. This book aims to develop an understanding of the basis and intent of the Code rules.

Like many codes, standards, and specifications, B31.3 can be difficult to understand and apply. There are endless cross references to explore during problem solving and the subject matter often overlaps several technical disciplines. B31.3 assumes that Code users have a good understanding of a broad range of subjects, but experience often shows the extent of understanding to be widely variable and restricted to a specific technical area. This book offers some insight into the basic technologies associated with design, materials, fabrication, testing, and examination of process piping systems.

B31.3 does not address all aspects of design, materials, fabrication, testing, and examination of process piping systems. Although the minimum requirements of the Code must be incorporated into a sound engineering design, the Code is not a substitute for sound engineering. A substantial amount of additional detail may be necessary to completely engineer and construct a process piping system, depending upon the piping scope and complexity. This book includes supplementary information that the Code does not specifically address. The intent of this information is generally to enhance the Code user's understanding of the broad scope of process piping system design, material selection, fabrication techniques, testing practices, and examination methods.

As an active member of the B31.3 committee since 1979, Glynn Woods has seen many questions from Code users asking for explanations of the Code's intent, position, and application. Likewise, Roy Baguley's international experiences as a metallurgical and welding engineer have involved the application of the Code in many different countries. These experiences have provided the practical engineering background needed to write this book. It was the challenge of the authors to make the reader feel more at ease with the use and application of the B31.3 Code and gain a greater insight into the Code.

Editor's Note

Chapters 1, 2, 3, 4, 8, and 9, constituting the "design" portion of this book, were written by Glynn Woods, while Chapters 5, 6, 7, and Appendix 1 constituting the "materials/welding/inspection" portion of this book, were written by Roy Baguley.

Note that the symbol "¶" precedes a Code paragraph referenced in the text of this book, for example, ¶304.1.2 refers to B31.3 paragraph 304.1.2.

TABLE OF CONTENTS

Chapter 1

INTRODUCTION

History of Piping and Vessel Codes

The realization of the need for codes did not become apparent until the invention of the steam engine. The first commercially successful steam engine was patented by Thomas Savery of England in 1698. The Savery engine, and the numerous improved engines which followed, marked the beginning of the industrial revolution. This new economical source of power was used to drive machines in factories and even enabled new and faster forms of transportation to be developed.

The boilers of these early steam engines were little more than tea kettles where direct heating of the boiler wall was the method used to generate the steam. These crude boilers were the beginning of pressure containment systems.

Boiler designers and constructors had to rely only on their acquired knowledge in producing boilers because there were no design and construction codes to guide them in their efforts to manufacture a safe operating steam boiler. Their knowledge was inadequate as evidenced by the numerous boiler explosions that occurred. A few of the more spectacular explosions will be mentioned.

On April 27, 1865, at the conclusion of the Civil War, 2,021 Union prisoners of war were released from Confederate prison camps at Vicksburg, Mississippi. Their transportation home was aboard the Mississippi River steamboat Sultana (Figure 1.1). Seven miles north of Memphis, the boilers of the Sultana exploded. The boat was totally destroyed; 1,547 of the passengers were killed. This event killed more than twice as many people as did the great San Francisco earthquake and fire of 1906.

In 1894, another spectacular explosion occurred in which 27 boilers out of a battery of 36 burst in rapid succession at a coal mine near Shamokin, Pennsylvania, totally destroying the entire facility and killing 6 people.

Boiler explosions continued to occur. In the ten-year period from 1895 to 1905, 3,612 boiler explosions were recorded, an average of one per day. The loss of life ran twice this rate - over 7,600 people were killed. In Brockton, Massachusetts on March 20, 1905, the R. B. Grover Shoe Company plant (Figure 1.2a before, Figure 1.2b after) was destroyed, killing 58 and injuring 117. A year later in Lynn, Massachusetts, a $500,000 loss from a night-time factory boiler explosion occurred injuring 3 people.

Figure 1.1 The Sultana.

The problem was clearly defined. Steam boilers, although a valuable source of power, were not safe. An engineering solution had to be found to protect against these disastrous explosions. This solution was first introduced in August 1907 by the state of Massachusetts with the establishment of the Board of Boiler Rules, the first effective boiler design legislation in the United States.

Other states followed with their own boiler rules: 1911, New York and Ohio, 1913, New Jersey, 1915, Indiana, 1916, Pennsylvania, 1917, California, Michigan, and Arkansas, 1918, Delaware and Oklahoma, 1920, Oregon, and so forth.

Figure 1.2a R.B. Grover Shoe Company. March 19, 1905, before explosion.

Figure 1.2b R.B. Grover Shoe Company. March 20, 1905, after explosion.

However, with all this legislation by the states, no two had the same rules. Great difficulties resulted in validating the inspection of boilers destined for out of state use. Even materials and welding procedures considered safe in one state were prohibited in another.

The American Society of Mechanical Engineers (ASME), already recognized as the foremost engineering organization in the United States, was urged by interested sections of its membership to formulate and recommend a uniform standard specification for design, construction, and operation of steam boilers and other pressure vessels.

On February 15, 1915, SECTION 1, POWER BOILERS, the first ASME boiler code, was submitted to council for ASME approval. Other code sections followed during the next eleven years:

Section III - Locomotive Boilers, 1921
Section V - Miniature Boilers, 1922
Section VI - Heating Boilers, 1923
Section II - Materials and Section VI Inspection, 1924
Section VIII - Unfired Pressure Vessels, 1925
Section VII - Care and Use of Boilers, 1926.

Figure 1.3 graphically illustrates the effectiveness of codes with their collective effort to present design rules and guidelines for designers and constructors to produce safe steam boilers. Here it can be seen there was a rapid decline in steam boiler explosions even as steam pressure steadily increased. Each of these code sections was written by committees of individuals with various areas of expertise in design, fabrication, and construction of boilers and pressure vessels. The committees' duty was to formulate safety rules and to interpret these rules for inquirers.

In 1934, an API-ASME code made its first appearance for large vessels operating at elevated temperatures and pressures. A second edition was released in 1936. However, the API-ASME Vessel Code was less conservative than the ASME Section VIII code that was established in 1925, nine years earlier. From 1935 to 1956, the members of the two code committees deliberated. The result was that the API-ASME code was abandoned and the ASME Boiler And Pressure Code Section VIII was adopted.

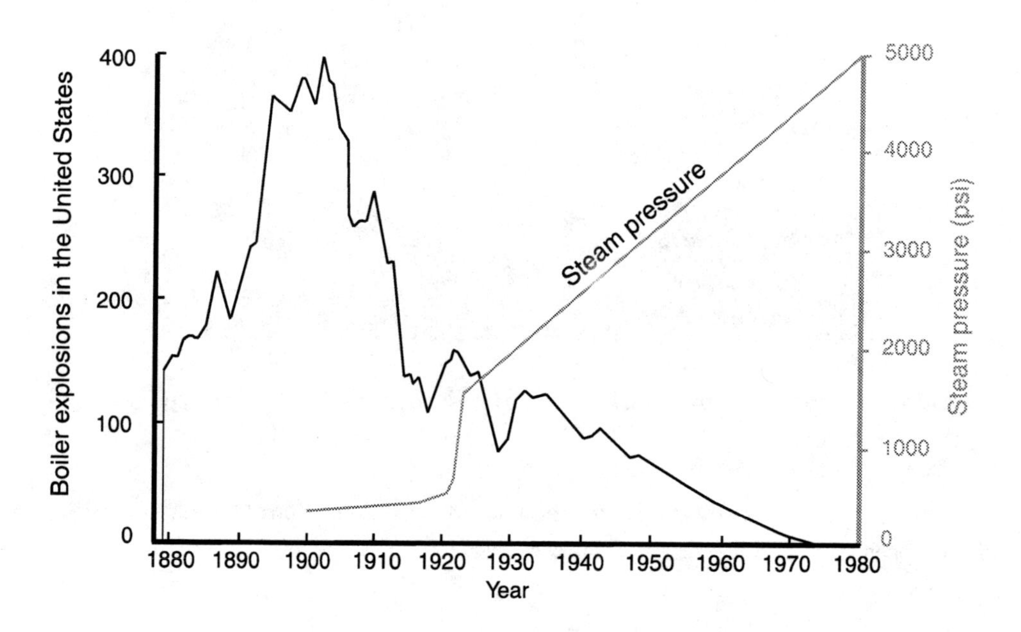

Figure 1.3 Effective application of ASME codes and standards resulted in a dramatic decline in boiler explosions.

The code for Pressure Piping has emerged much the same way as the pressure vessel code. To meet the need for a national pressure piping code, the American Standards Association (ASA) initiated PROJECT B31 in March 1926, at the request of ASME and with ASME being the sole sponsor. Because of the wide field involved, Section Committee B31 comprised some forty different engineering societies, industries, government bureaus, institutions, and trade associations. The first edition of the B31 Code was published in 1935 as the American Tentative Standard Code for Pressure Piping.

To keep the Code current with developments in piping design and all related disciplines, revisions, supplements, and new editions of the Code were published as follows:

B31.1 - 1942 American Standard Code for Pressure Piping
B31.1A - 1944 Supplement 1
B31.1B - 1947 Supplement 2
B31.1 - 1951 American Standard Code for Pressure Piping
B31.1A - 1953 Supplement 1 to B31.1 - 1951
B31.1 - 1955 American Standard Code for Pressure Piping

The first edition of the Petroleum Refinery Piping Code was published as ASA B31.3-1959 and superseded Section 3 of B31.1 - 1955. Two subsequent editions were published, ASA B31.3 - 1962 and ASA B31.3 - 1966.

In 1967, the American Standards Association was reorganized, and its name was changed to the United States of America Standards Institute. In 1969, the Institute changed its name to American National Standards Institute (ANSI).

In 1973, a new Petroleum Refinery Piping Code designated as ANSI B31.3 - 1973 was published.

A code for Chemical Plant Piping, designated ANSI B31.6, sponsored by the Chemical Manufacturers Association was in preparation but not ready for issue in 1974. At this time, in response to an inquiry to the ASME Code Piping Committee, Code Case 49 was issued, and instructed designers to use B31.3 requirements for chemical plant piping.

> **Code Case 49**, Code Section to Be Used for Chemical-Industry Piping Inquiry.
> **Inquiry**: Is there a Code Section of ASA B31 (Code for Pressure Piping) by which chemical process industry piping may be designed, fabricated, inspected, and tested?
> **Reply**: It is the opinion of the Committee that until such time as an ASA Pressure Piping Code Section specifically applying to chemical process piping has been published, chemical process piping may be designed, fabricated, inspected, and tested in accordance with the requirements of ASA B31.3, Petroleum Refinery Piping.

Rather than publish two code sections, it was then decided to combine the requirements of B31.3 and B31.6 into a new edition designated ANSI B31.3, with the title changed to Chemical Plant and Petroleum Refinery Piping. This edition first appeared in 1976.

In 1978, the American Standards Committee B31 was reorganized as the ASME Code for Pressure Piping, under the procedures developed by ASME and accredited by ANSI.

The 1980 edition of the Chemical Plant and Petroleum Refinery Piping Code appeared as ANSI/ASME B31.3.

In July, 1981, a code section written for Cryogenic Piping, ANSI/ASME B31.10, followed the path taken by B31.6 in 1976. It too was dissolved and the rules developed for cryogenic piping were incorporated into the guidelines of B31.3. Changes in B31.3 are occurring to better accommodate cryogenic piping. The decision was made not to rename the code to reflect cryogenic piping coverage; the current code is ASME B31.3.

In March 1996, a new base Code was published with its name changed to ASME B31.3 Process Piping. It is with this edition that B31.3 will start adopting the metric system of measurement. The planned 1999 edition of B31.3 is to be 100% metric units.

Scope

As seen in the preceding section, codes originated in response to numerous boiler explosions resulting from unsafe design and construction practices. It is not surprising then that the primary goal of codes is safety. This goal is achieved through putting into place a set of engineering requirements deemed necessary for safe design and construction of piping systems. In addition, prohibitions and warnings about unsafe designs and practices are also included.

The B31.3 Code also provides:

1. A list of acceptable piping materials with their allowable stress at various temperatures and numerous notes providing additional information on the use of each material.
2. A tabulation of standards which include acceptable components for use in B31.3 piping systems such as:
   a) ANSI B16.5, which covers the dimensions, materials of construction, and the pressure-temperature limitations of the common types of flanges found in refinery piping.
   b) ANSI B16.9, another dimensional standard for butt-welded fittings such as tees, crosses, elbows, reducers, weld caps, and lap joint stub ends. B16.9 fittings must also be capable of retaining a minimum calculable pressure.
   c) ANSI B16.11, another dimensional standard for socket-weld and threaded tees, couplings, and half-couplings. This standard also has a minimum pressure requirement.
   These are only a few of the more than 80 listed standards.
3. Guidance in determining safe piping stress levels and design life.
4. Weld examination requirements for gaging the structural integrity of welds.
5. Pressure test requirements for piping systems before plant start-up.

With the above in mind, it might be assumed that the B31.3 Code is a designer's handbook. This belief could not be further from the truth. *The Code is not a design handbook and does not eliminate the need for the designer or for competent engineering judgment* [¶B31.3 Introduction]. The Code provides only a means to analyze the design of a piping system, by providing simplified equations to determine the stress levels, wall thickness, or the design adequacy of components, and acceptance criteria for examination. *The Code does not provide any instruction on how to design anything.*

The Code's approach to calculating stress levels and assuring safety in piping is a simplified one [¶B31.3 Introduction]. Codes would be of little use if the equations specified were very complicated and difficult to use. Codes would find little acceptance if their techniques and procedures were beyond the understanding of the piping engineer. This is not to say, however, that designers who are capable of applying a more rigorous analysis should be restricted to this simplified approach. In fact, such designers who are capable of applying a more rigorous analysis have the latitude to do so provided they can demonstrate the validity of their approach.

The choice of a code to comply with for a new piping system, lies for the most part, with the plant owner. With the exception of a few states and Canadian provinces, B31.3 is not mandated by law. The states and provinces that have made this Code mandatory as of mid 1985 are:

- Colorado
- Connecticut
- Kentucky
- Ohio
- Washington
- Washington, D C
- Prince Edward Island
- Alberta
- British Columbia
- Manitoba
- Newfoundland
- Nova Scotia
- Ontario

On occasion, the plant owner may have to decide which code section to use for a particular plant. Two different codes, for example, may have overlapping coverage where either may be suitable. Cogeneration plants within a refinery, for example, could be designed either to B31.1 or B31.3. The answers to two questions could be helpful in selecting the governing code section:

1. How long do you want the plant to last?

2. What reliability do you want the plant to have?

Plants designed to B31.3 generally have a life of about 20 to 30 years. Plants designed to B31.1, on the other hand, may be expected to have a plant life of about 40 years. The difference between these two codes is the factor of safety in the lower to moderate design temperature range. B31.3 uses a 3 to 1 factor of safety, where B31.1 has a 4 to 1 factor. This factor can reflect differences in plant cost. For example, the same design conditions for a B31.1 piping system may require schedule 80 pipe wall thickness, while a B31.3 system on the other hand, may require only schedule 40 pipe wall thickness.

Plant reliability issues center on the effect of an unplanned shutdown. Loss of power to homes on a cold winter night is an example of a reason to have very high plant reliability. Here, the safety of the general public is affected. If a chemical plant is forced off stream for one reason or another, very few people are affected. A lesser reliability can be tolerated.

The types of plants for which B31.3 is usually selected are: installations handling fluids including fluidized solids; raw, intermediate, or finished chemicals; oil; petroleum products; gas; steam; air; and refrigerants (not already covered by B31.5). These installations are similar to refining or processing plants in their property requirements and include:

- chemical plants
- loading terminals
- bulk plants
- tank farms
- food processing
- pulp & paper mills
- off-shore platforms
- petroleum refineries
- natural gas processing plants
- compounding plants
- steel mills
- beer breweries
- nuclear fuel reprocessing plants

Definitions

In applying the Code, the designer must have a working knowledge and understanding of several key terms and conditions. This will greatly assist the designer in applying the *intent* of the Code. A few of the fundamental terms and conditions are defined below.

Principal Axis and Stress

The analysis of piping loaded by pressure, weight, and thermal expansion can appear to be rather complicated and difficult to accomplish. This complexity will be greatly simplified when the analyst has an understanding of the *Principal Axis System*.

Consider a cube removed from a stressed section of a pipe wall that is acted upon by forces from several direction. What is the remaining safety from failure or over strain in the cube?

To answer this question, simply calculate the stress in the cube and compare it to some allowable stress limit.

Stress is defined as the ratio of force to area. To find the stress in the cube, construct a three-dimensional, mutually perpendicular *principal axis* system, with each axis perpendicular to the face of the cube it penetrates (Figure 1.4). The origin of the principal axis system is at the center of the cube.

Each force acting on the cube can be trigonometrically reduced to force components, represented by vectors, acting along each of the principal axes. The resultant of the component of each force acting on the face of the cube, divided by the area of the cube face is called the *principal stress*. The principal stress acting along the center line of the pipe is called a *longitudinal principal stress*. This stress is caused by longitudinal bending, axial force loading, or pressure.

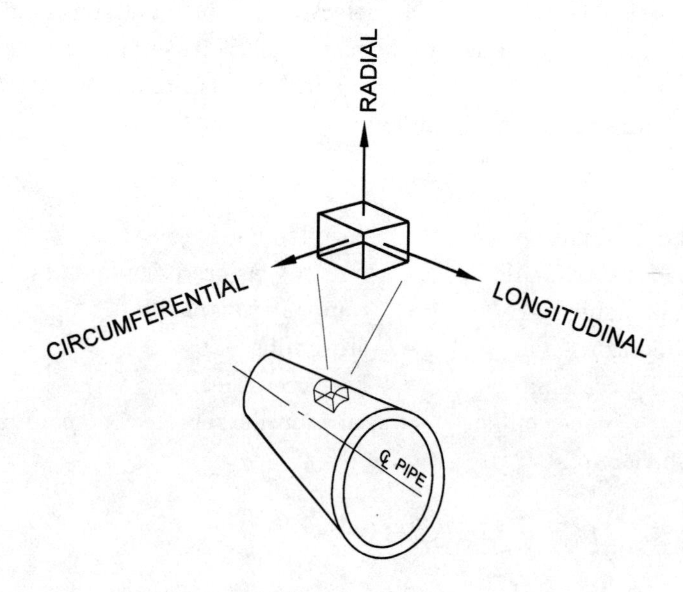

Figure 1.4 Principal axis system.

Radial principal stress acts on a line from the center of pipe radially through the pipe wall. This stress is a compressive stress acting on the pipe inside diameter caused by internal pressure or a tensile stress caused by vacuum pressure.

Circumferential principal stress, sometimes called *hoop* or *tangential stress,* acts on a line perpendicular to the longitudinal and radial stresses. This stress tends to separate the pipe wall in the circumferential direction and is caused by internal pressure.

When two or more principal stresses act at a point on a pipe, a *shear stress* will be generated. One example of a shear stress would be at a pipe support where radial stress caused by the supporting member acts in combination with the longitudinal bending caused by the pipe overhang.

In the following discussion, principal stresses are used to present failure theories which are based on the yield strength of materials. Figure 1.5 is a typical stress-strain representation of the yield strength of a typical annealed steel.

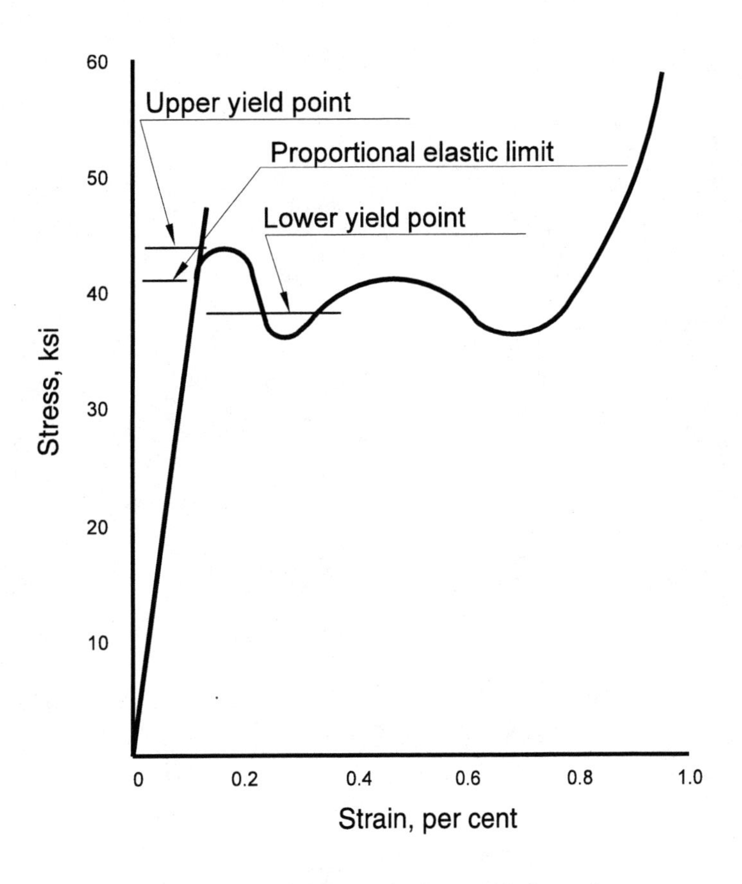

Figure 1.5 Yielding of an annealed steel.

Failure Theories

The Code presents equations for determining the stress levels in a piping system and also provides allowable stress limits for comparison [¶302.3.5, 319.4.4 for example]. These Code equations are based on failure theories. A review of these theories is helpful in gaining a greater understanding of the intent of the Code. These theories are the *maximum principal stress failure theory* and the *maximum shear stress failure theory* (TRESCA).

The maximum principal stress failure theory states that when any one of the three mutually-perpendicular principal stresses exceed the yield strength of the material at temperature, failure will occur. An example of the application of the maximum principal stress theory follows:

Calculate the principal stresses in a 14 in. outside diameter, 0.375 in. wall thickness pipe operating at 1200 psig internal pressure.

Solution:

Longitudinal Principal Stress, (LPS):

$$LPS = \frac{PD}{4T} = \frac{1200 \times 14}{4 \times 0.375} = 11{,}200 \text{ psi}$$

Circumferential Principal Stress, (CPS):

$$CPS = \frac{PD}{2T} = \frac{1200 \times 14}{2 \times 0.375} = 22{,}400 \text{ psi}$$

Radial Principal Stress, (RPS):

RPS = P
RPS = -1200 psi on the inside surface and
RPS = 0 on the outside surface of the pipe.

Applying the maximum principal stress failure theory to this piping condition, the circumferential principal stress would be the only stress of any concern. To protect against failure, a wall thickness must be selected that will produce a hoop stress below the yield strength of the piping material at the temperature of the pressured condition.

The maximum shear failure theory is an arithmetic average of the largest minus the smallest principal stress. (Tension-producing principal stresses are positive, and compression principal stresses are negative.) Earlier, shear stress was defined as two or more principal stresses acting at the same point in the pipe.

The example used in the maximum principal stress theory can be used to explain the maximum shear failure theory.

The three principal stresses acting at the same point in the previous example were:
LPS = 11,200 psi; CPS = 22,400 psi; RPS = -1,200 psi on the inside.

The maximum shear (MS) for this example would be;

$$MS = \frac{CPS - RPS}{2} = \frac{22{,}400 - (-1{,}200)}{2} = 11{,}800 \text{ psi}$$

The maximum shear failure theory states that when the maximum shear stress exceeds one-half of the yield strength of the material at temperature, failure will occur. In the example above, this system will be safe as long as the yield strength at temperature is above 23,600 psi.

Stress Categories

The B31.3 Code provides design guidance for *primary* and *secondary stresses*. Although the Code user will not find these terms listed or discussed in the Code text, a fundamental understanding of these terms is essential in applying the Code equations to calculate the stress levels in piping systems and to compare these calculated stresses to the Code allowable stresses.

A primary stress is a principal stress, shear stress or bending stress generated by imposed loadings which are necessary to satisfy the simple laws of equilibrium of internal and external forces and moments. The basic characteristic of a primary stress is that it is not self-limiting. As long as the load is applied, the stress will be present and will not diminish with time or as deformation takes place. The failure mode of a primary stress is gross deformation progressing to rupture.

Examples of a primary stress are circumferential stresses due to internal pressure and longitudinal bending stresses due to gravity (see Figure 1.6, and Figure 1.7, top).

Figure 1.6 Primary stress failure.

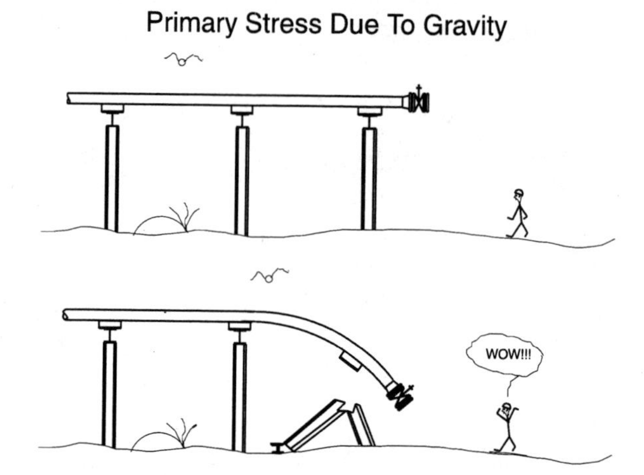

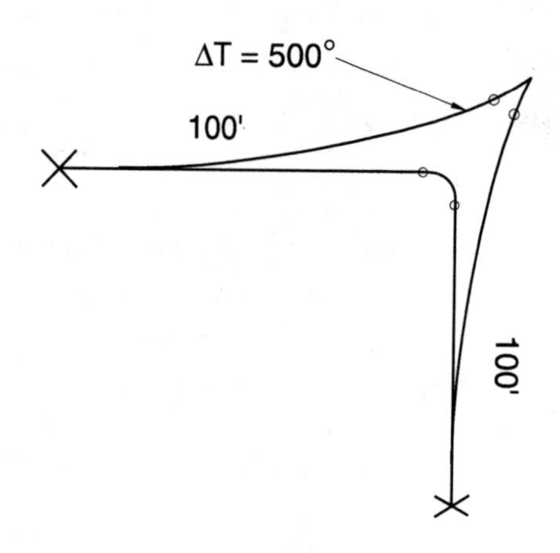

Figure 1.7 Primary and secondary stresses.

A secondary stress is a principal stress, shear stress or bending stress caused by structural restraints such as flexibility controls or by constraint of the pipe itself. The basic characteristic of a secondary stress is that it is self-limiting. As this stress condition develops in a piping system, local yielding will occur, thus reducing these stresses.

An example of a secondary stress is the bending of an elbow that joins two lengths of pipe subjected to a temperature increase (see Figure 1.7, bottom). As the piping reaches its operating temperature, the bending strain (stress) in the elbow will reach a maximum magnitude and the strain (stress) will stabilize. Thus, the condition causing the stress to increase stops. The elbow, now in this new higher stressed state, will in time, experience local yielding or deformation which will reduce the stresses in the elbow.

One can see from this example that secondary stresses in a piping system are associated with cyclic conditions such as temperature increase or decrease, as the plant starts up or shuts down. The failure mode of a secondary stress is fatigue crack initiation and propagation through the pressure boundary resulting in a leak (see Figure 1.8).

Figure 1.8 Secondary stress (fatigue) failure crack.

Peak Stress, although not specifically discussed in B31.3, is worthy of adding to the list of definitions. Peak stresses are those stresses which are caused by local discontinuities or abrupt changes in a pipe wall thickness when the pipe is subjected to a primary or secondary stress loading. Peak stresses are stress concentration points which can cause crack initiation contributing to a fatigue failure.

Definition and Basis for Allowable Stress, S_c and S_h

An understanding of the term *allowable stress* is essential when discussing the piping codes. There is a Code allowable stress for pressure design, others for pure compression, shear and tension [¶302.3], and still another for thermal displacement stresses calculated from equations using the terms, "S_c" and "S_h". To increase the understanding of the use of these allowable stress equations and terms, a review of the basis of these two terms will be beneficial.

The allowable stress for a piping system or a piping component material is based on a function of the yield or tensile strength of the material at cold to moderate temperatures, or is based on creep rates or stress for rupture in elevated temperature service.

The term "S_c" is the allowable stress for a material at the *cold* condition, which includes cryogenic service, or ambient installed temperature for elevated temperature service.

"S_h" is the allowable stress for the material in the *hot* operating condition, which would be the design temperature for elevated temperature service or ambient for cold or cryogenic service.

The values for S_c and S_h are tabulated in Appendix A Table A-1 of the B31.3 Code for most materials used in refinery piping service. These values are the same as those of the BPV Code Section VIII, Division 2 (as published in ASME Section II, Part D), using a 3 to 1 factor of safety for temperatures below the creep range. Generally, creep range temperatures are those above 700 to 800°F. Each Table A-1 value for temperatures below the creep range is the lowest of the following conditions:

1. One-third of the specified minimum tensile strength at room temperature;
2. One-third of the tensile strength at temperature;
3. Two-thirds of the specified minimum yield strength at room temperature;
4. Two-thirds of the yield strength at temperature. (For austenitic stainless steels and certain nickel alloys, this value may be as large as 90% of the yield strength at temperature [¶302.3.2(e)].)

Figure 1.9 illustrates how the allowable stress decreases for two materials as temperature is increased. Notice for the ASTM A 53 Gr. B material how the decrease in allowable stress becomes non-linear as temperature exceeds 700°F. The stainless material displays this non-linear characteristic as the temperature exceeds 900°F. This non-linearity is representative of the Code allowable stress limitations imposed on these materials at creep range temperatures.

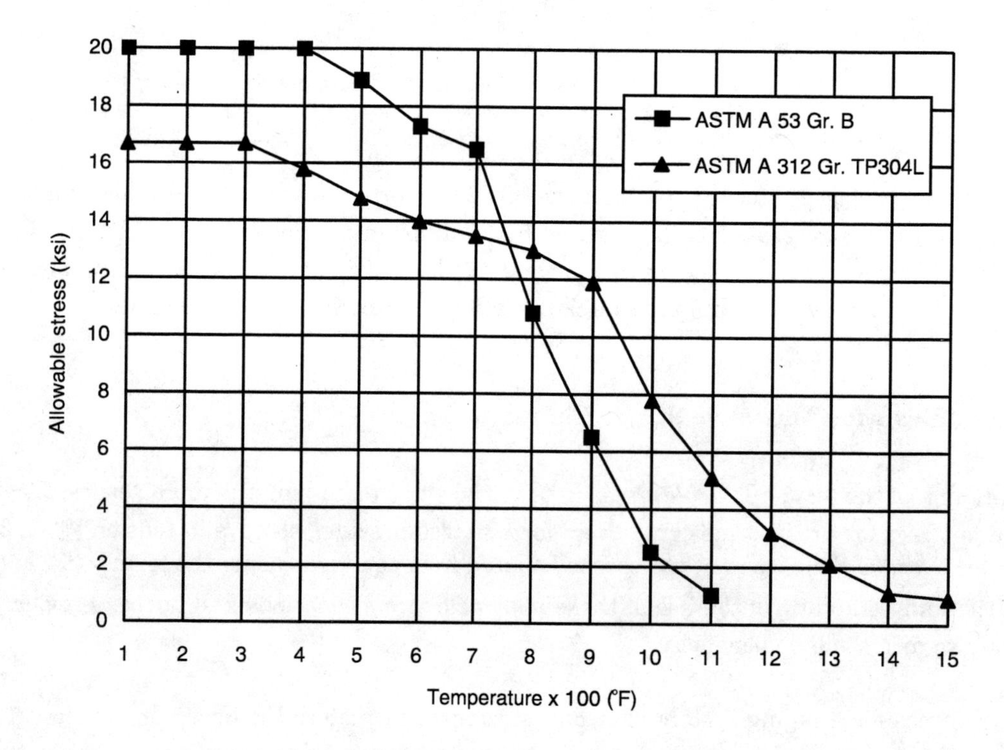

Figure 1.9 Temperature effect on material allowable stress.

Chapter 2

PRESSURE DESIGN OF PIPING & PIPING COMPONENTS

Design Conditions

An essential part of every piping system design effort is the establishment of the design conditions for each process. Once they are established, these conditions become the basis of that system's design. The key components of the design conditions are the design pressure and the design temperature.

Design Pressure and Temperature

Design pressure is defined as the most severe sustained pressure which results in the greatest component thickness and the highest component pressure rating. It shall not be less than the pressure at the most severe condition of coincident internal or external pressure and maximum or minimum temperature expected during service [¶301.2].

Design temperature is defined as the sustained pipe metal temperature representing the most severe conditions of coincident pressure and temperature [¶301.3]. B31.3 provides guidance on how to determine the pipe metal temperature for hot or cold pipe in ¶301.3.2.

Designers must be aware that more than one design condition may exist in any single piping system. One design condition may establish the pipe wall thickness and another may establish the component rating, such as for flanges.

Once the design pressure and temperature have been established for a system, the question could be asked: Can these condition ever be exceeded? The answer is yes, they can be exceeded. In the normal operation of a refinery or chemical plant, there is a need, on occasion, for catalyst regeneration, steam-out or other short term conditions that may cause temperature-pressure variations above design. Rather than base the design pressure and temperature on these short term operations, the Code provides conditions to permit these variations to occur without becoming the basis of design.

A review of ¶302.2.4, Allowances for Pressure and Temperature Variations, Metallic Piping, reveals these conditions for variations. Therein, the Code sets the first two allowable stresses for design:

1. The nominal pressure stress (hoop stress), shall not exceed the yield strength of the material at temperature.
2. The sum of the longitudinal stresses due to pressure, weight, and other sustained loadings plus stresses produced by occasional loads, such as wind or earthquake, may be as high as 1.33 times the hot allowable stress, S_h, for a hot operating system [¶302.3.6].

Before continuing on, let's apply what has been covered in order to understand the basis of the limits the Code places on these two stresses.

Pressure stress in the first condition above is the circumferential (principal) stress or hoop stress defined earlier. The stress limit of the yield strength at temperature is simply a restatement of the maximum principal stress failure theory. If indeed, the hoop stress exceeded the yield strength at temperature, a primary stress failure would occur.

The second stress condition, the longitudinal stress caused by pressure and weight, is a principal stress and pressure, weight and other sustained loadings are primary stress loadings. The allowable stress, S_h, is defined earlier in Chapter 1 as a stress limit value that will not exceed a series of conditions, one of which was ⅔ yield at temperature. Applying this ⅔ yield stress condition with the 1.33 S_h stress limit, we find again, a direct application of the maximum principal stress failure theory. That is, longitudinal principal stress must be less than 1.33 x ⅔ yield strength at temperature [¶302.3.6], the product of which results in a limit of about 90% yield. Again, the primary stress is less than yield at temperature. (Some factor of safety is included in this equation to account for the simplified technique of combining these stresses.)

Continuing to study the conditions for pressure-temperature variations, we find one of the most misinterpreted and misapplied statements of the Code. It is in ¶302.2.4(1) (where the allowable stress for pressure design is S_h):

> ... it is permissible to exceed the pressure rating or the allowable stress for pressure design at the temperature of the increased condition by not more than:
>
> a) 33% for no more than 10 hours at any one time and no more than 100 hours/year; or
> b) 20% for no more than 50 hours at one time and no more than 500 hours/year.

What is the Code saying? What is the basis of these time dependent stress or rating limits?

Allowing the pressure rating of components, such as flanges, to be exceeded by as much as 33% will permit the stresses to approach yield in the flange without causing a genuine concern for over stress. Flange rating procedures will be discussed later in this chapter. Caution must be exercised when the allowable stress for pressure design is based on 90% yield at temperature as in the case of austenitic stainless steels used in higher temperature service. Here, pressure stresses which exceed S_h by 33% can cause deformation and leakage in the flange. For these stainless steels, the pressure design allowable stress should be based on 75% of S_h from B31.3 Table A-1 or on ⅔ of the yield strength of the material listed in ASME Section II Part D [¶302.3.2(e)].

The Code statement permitting the allowable stress for pressure design to be exceeded has confused many designers. The allowable stress for pressure design is S_h, the basic allowable stress of the material at the hot temperature. Often this statement is mistakenly used to increase the allowable stress range, "S_A", the allowable stress for displacement stresses, "S_E" by 33%. This is not the intent of the Code.

It is interesting that ¶302.2.4 includes time dependent stress limits. What is the basis of these stress limits?

These stress limits are based on the *use-fraction sum rule*, which states:

$$\Sigma \frac{t(i)}{t(ri)} \leq 1.0$$

where

t(i) = the total lifetime in hours associated with a given pressure P(i) and/or temperature T(i).

t(ri) = the allowable time in hours at a given stress corresponding to a given pressure P(i) and temperature T(i). Such t(ri) values are obtained by entering the stress-to-rupture curve for the particular material at a stress value equal to the calculated stress S(i) divided by 0.8. This action adds a 25% factor of safety by escalating the calculated stress. Designers may select another factor of safety depending on their particular conditions. The stress-to-rupture curves, found in ASME Code Case N47, do not have a factor of safety built in. They are failure curves, not design curves.

The use fraction sum rule is illustrated by the following example:

Assume that the sustained load (primary) stresses, "S_L", in an elbow caused by pressure, weight, and other sustained loadings is 5,000 psi at 1100°F and 600 psig pressure. Further assume that the process requires a short time pressure-temperature variation above normal operating as shown in Table 2.1. Applying the use fraction sum rule, is the elbow over stressed? The elbow material is ASTM A 358 Type 304; the plant life is 10 years.

Table 2.1 Pressure-Temperature Operating Status

Mode	**P(i) (psig)**	**T(i) (°F)**	**Frequency and hours per event**	**S(i) (psi)**	**t(ri), Total Time, (hr.)**
Normal Operations	600	1100	Continuous	5,000	100,000
Pressure Surge	700	1100	12 events per year, 40 hours duration	5,300	4,800
Temperature Surge	600	1200	10 events per year, 10 hour duration	5,400	1,000
Pressure-Temperature	750	1250	3 events per year, 10 hour duration	5,800	300

The question is, are normal operating "P(i)" and "T(i)" to be the design pressure and temperature or are they design conditions to be replaced by a variation?

The allowable stress limit for S_L based on normal operating conditions is the hot allowable, S_h, which is: S_h = 9,700 psi (from B31.3 Table A-1, for ASTM A 358 Type 304 at 1100°F). Note that S_h is not exceeded, but should the design conditions be changed? Solution: Construct the time fraction table (Table 2.2) using the allowable hour at stress vs. temperature graph shown in Figure 2.1.

Table 2.2 Use Fraction Table

Mode	S(i)/0.8 (psi)	T(i) (°F)	t(i) (hours)	t(ri) (hours)	t(i)/t(ri)
Normal operations	6,250	1100	100,000	200,000	0.50
P Surge	6,625	1100	4,800	100,000	0.05
T Surge	6,750	1200	1,000	10,000	0.10
T-P Surge	7,250	1250	300	2,000	0.15
				Use Fraction Sum =	**0.80**

The use fraction sum is less than 1.0; therefore, pressure-temperature variations are within the time and calculated stresses are within the time-stress criteria as specified in ¶302.2.4 of the Code. The elbow will not be over stressed and the design pressure and temperature will not have to be ~~increased.~~ DECREASED

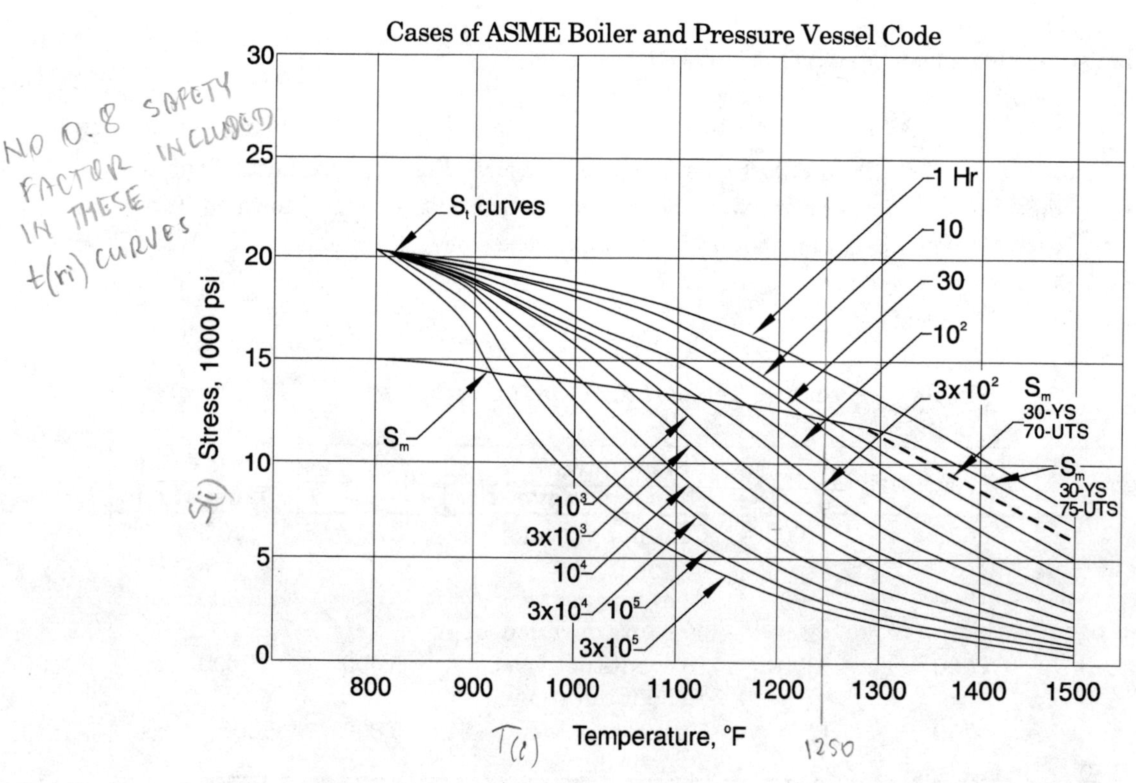

Figure 2.1 Timed allowable stress per temperature for Type 304 stainless steel.

Considerations of Design

In addition to the design temperature and pressure, there are several other considerations of design that must be addressed to ensure a safe operating piping system. The Code lists several of these considerations beginning with ¶301.4 and provides a good explanation of each. However, there is one consideration for which the explanation should be expanded. That is the discussion on vibration [¶301.5.4].

Vibration

The guidance presented in the Code for checking cyclic stress levels is based on low cycle, high stress. In a vibrating system, the stress concern is high cycle, low stress. A clarification of what is meant by high and low cycle is in order.

The Code allowable stress range for cyclic stresses, S_A [¶302.3.5], is based in part, on the number of thermal or equivalent cycles the system will experience in the plant life. Table 302.3.5 of the Code tabulates a factor used to determine S_A, called the *stress-range reduction factors* ("f"). f ranges from 1.0 for 7,000 cycles or less (7,000 cycles is about one cycle per day for 20 years), to 0.3 for cycles up to 2,000,000. The intent of the Code is to provide an allowable stress reduction factor for the secondary stress cycles expected in the lifetime of the plant.

A vibrating piping system (see Figure 2.2) can easily experience more than 500,000 stress cycles in a single day. Clearly, the stress range reduction factor-allowable stress range philosophy is not applicable for vibrating piping systems. The Code does not address high cycle - low stress piping life in vibrating systems.

How then does one analyze a vibrating pipe? One answer to this question is to:

1. Calculate the stress level, S_E, caused by the displacement in the vibrating pipe [¶319.4.4].
2. Estimate the number of vibrating cycles expected in the life of the plant.
3. Enter the ASME BPV code design fatigue curves for the pipe material to determine if the stress-cycle intersection point will be below the fatigue curve. If it is, the vibrating system should last the design plant life.

Design fatigue curves are presented in Appendix 5-Mandatory of the ASME BPV Code Section VIII Division 2. The fatigue curve for carbon steel operating at temperature service not over 700°F is presented in Figure 2.3. As an example, consider the use of this guideline to determine the cycle life of a carbon steel elbow where S_E has been calculated to be 30,000 psi.

Intersecting a line from 30,000 psi to the ultimate tensile strength (UTS) < 80 line gives a cycle life of about 35,000 cycles. Typical vibrating stresses would be in the range of 1,000 to 2,000 psi. Another graph covering that stress and cycle range would have to be selected to address cycle life of that lower stress range.

Figure 2.2 Reciprocating compressors are sources of vibrating pipe.

ASME Section VIII - Division 2, 1992. APPENDIX 5 - Fig. 5-110 - Mandatory

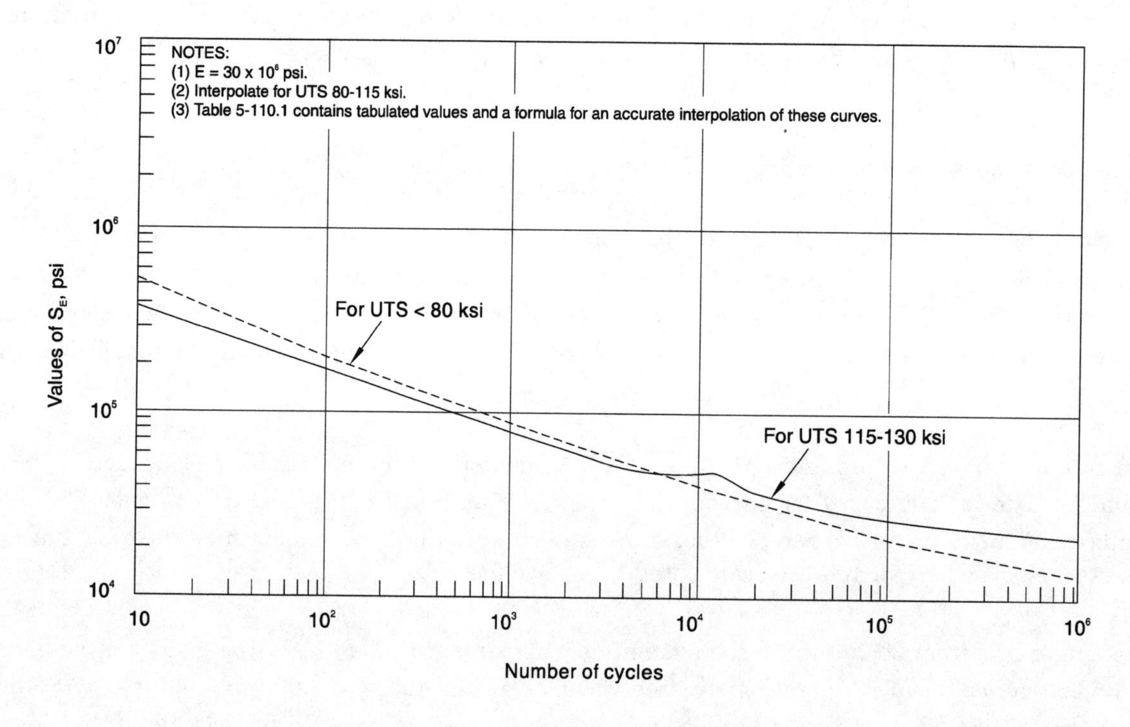

Figure 2.3 Design fatigue curves for carbon, low alloy, series 4XX, high alloy steels and high tensile steels for temperatures not exceeding 700°F.

Water Hammer [¶301.5]

Water hammer and pressure surge are piping systems design considerations where the designer can find assistance in the AWWA Steel Pipe Manual (AWWA M11) in predicting the pressure rise in a liquid system caused by rapid valve closure. An example follows:

The pressure rise ("P") for instantaneous valve closure is directly proportional to the fluid-velocity ("V") cutoff and to the magnitude of the surge wave velocity ("a") and is independent of the length of the pipe.

$$P = \frac{aWV}{144g}$$

$$a = \frac{12}{\sqrt{(W/g) \times ((1/k) + (d/Ee))}} \text{ fps}$$

where

a = wave velocity (fps)
P = pressure rise above normal (psi)
V = velocity of flow (fps)
W = weight of fluid (lb/cu ft)
k = bulk modulus of elasticity of liquid (psi)
E = Young's modulus of elasticity of pipe material (psi)
d = inside diameter of pipe (in.)
e = thickness of pipe wall (in.)
g = acceleration due to gravity (32.2 fps/sec)

For steel pipe,

$$a = \frac{4660}{\sqrt{(1 + (d/100e))}} \text{ fps}$$

For example, a rapid closing check valve closes in a 36 in. OD, 0.375 in. wall thickness pipe with a water velocity of 4 fps (k = 294,000 psi, E = 29,000,000 psi, and W = 62.4 lb/cu ft). What is the instantaneous pressure rise above operating pressure?

a = 3345 fps and

P = 180 psi pressure rise above normal.

This pressure rise acting at the closed valve in this piping system can exert a force equal to pressure times the cross sectional area of the pipe or about 175,665 pounds, which can cause an unrestrained pipe to move from its normal position.

Piping Design

Plant piping design is an essential part of successful plant operation. Many decisions must be made in the design phase to achieve this successful operation, including:

- required fluid quantity to or from a process
- the optimum pressure-temperature for the process
- piping material selection
- insulation selection
- stress and nozzle load determination
- pipe support scheme.

The Code provides minimal assistance with any of these decisions (the Code is not a design manual). However, the Code does address material suitability with respect to temperature. Pipe material selection for a particular fluid service based on the material reactivity, corrosion or erosion rates in the fluid service is not within the scope of the Code.

Appendix A, Table A-1 of the Code is a list recognized pressure-containing piping materials. By reviewing this table, the designer can determine: whether or not the material selected for service is "Code-recognized" (listed), the allowable stress (S_c or S_h) of the material for the process temperature, and if the Code offers any special considerations for use of the material. As an example of this Code assistance on material selection, consider the following:

ASTM A 53 Gr. B has been selected for service in the following process conditions,
T = 850°F; P = 600 psig.

Question: Is this a suitable material for these conditions?

Appendix A Table A-1 lists ASTM A 53 Gr. B, so it is a "Code recognized" material. There is a listing of an S_h, that satisfies another requirement in the check list. Finally, special considerations for use are covered in Note (57) of Table A-1, which states:

> Conversions of carbides to graphite may occur after prolonged exposure to temperatures over 800°F.

Therefore, this material is not suitable for normal operations above 800°F. Designers must be familiar with the behavior of the selected piping material when in contact with the service fluid, including its thermal expansion rate and any limitations the Code places on the material, including the notes of Table A-1.

Wall Thickness for Internal Pressure

The Code can also assist the designer in determining adequate pipe wall thickness for a given material and design conditions, as follows:

1. Calculate the pressure design thickness, "t".
2. Add the mechanical, corrosion, and erosion allowances, "c", to obtain the thickness, "t(m)" = t + c.
3. Add mill tolerance to t(m) then select the next commercially available schedule wall thickness.

This same procedure is used for designing non-scheduled pipe such as ID or OD controlled minimum wall pipe. This piping is usually found in higher temperature and pressure systems, such as power piping, B31.1 systems.

B31.3 offers four wall thickness equations for determining the pressure design thickness, t [¶304.1.2]. Any one of the four can be used provided t is less than D/6. If t is greater than D/6, the equations in Chapter IX on high pressure piping should be considered.

An example of the application of Code rules for wall thickness determination is:

Find the wall thickness of an 8.625 in. OD pipe for the conditions T = 500°F; P = 850 psig; c = 0.063 in. (corrosion-erosion allowance); material = ASTM A 53 Gr. B, ERW.

Solution: Of the four equations presented in ¶304.1.2 of the Code, equation (3a) will be selected to calculate the appropriate wall thickness. (The equation selection is the designer's choice. For low to moderately high temperatures, all OD wall thickness equations provided in ¶304.1.2 will produce about the same results.)

$$t = \frac{P \times D}{2\,(SE + PY)}$$

where

P = internal design gage pressure
D = pipe outside diameter
S = S_h taken from Appendix A
E = quality factor from Table A-1B
Y = stress-temperature compensating factor (nondimensional) taken from Table 304.1.1

Before continuing on, a word of explanation of the "E" and "Y" factors is in order. The E factor is an "allowable pressure stress penalty" based on the method of manufacture of the pipe. It reflects the quality of the longitudinal weld in seam-welded pipe and will have a value ranging from 0.6 for furnace butt welded (FBW) to 1.0 for seamless pipe, (SMLS). This factor is a carry-over from the days when pipe was manufactured using rivets or other means to make the seam joint (see Figure 2.4).

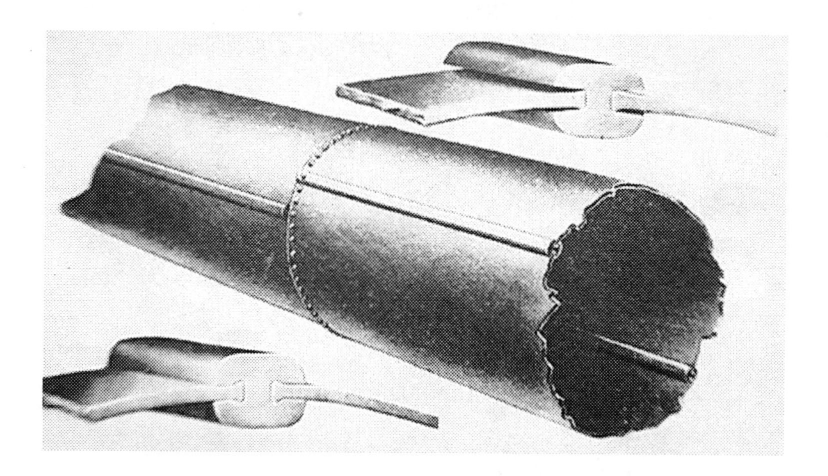

Figure 2.4 Longitudinal joint in early pipe.

The E factor for seam-welded pipe can be improved: increasing this factor from 0.8 to 1.0 for example, by supplementary nondestructive examination (NDE) presented in ¶302.3.4 of the Code. The E factor is applicable only for calculating the wall thickness of pipe and other pressure containing components.

The Y factor is included to account for the non-linear reduction in allowable stress at design temperatures above 900°F, see Figure 1.9.

Continuing with the solution to the wall thickness problem:

P = 850 psig
D = 8.625 inches
S = 18,900 psi at 500°F (from Table A-1, S_h)
E = 0.85 (Table A-1B)
Y = 0.4 (Table 304.1.1)

then $$t = \frac{8.625 \times 850}{2(18900 \times 0.85 + 0.4 \times 850)} = 0.223 \text{ in.}$$

and t(m) = 0.223 + 0.063 + 0.010 (mill under-run tolerance on rolled plate)
t(m) = 0.296 in.

The next commercially available pipe wall is schedule 40, with a nominal wall thickness of 0.322 in., which should be used for these stated conditions.

Using the term *schedule* to define wall thickness of pipe was first approved by ASME in 1934. The procedure for determining the schedule of pipe for a given service is,

$$\text{Schedule} = \frac{1000 \times \text{Pressure}}{\text{Sh}}$$

As example of the application of this equation, find the schedule of ASTM A 53 Gr. B pipe for 500°F, 1200 psig service:

Solution: S_h = 18,900 psi, then

$$\text{Schedule} = \frac{1000 \times 1200}{18900} = 63.5.$$

A schedule 80 pipe is the next commercially available schedule.

This example concerning schedule calculation is presented for information only and is not recommended for determining pipe wall thickness.

Figure 2.5 presents a comparison of the three B31.3 wall thickness equations for a particular pipe outside diameter and material at pressure as a function of increasing temperature. The following pages provide excerpts from ASTM Standards to be used in determining mill under-run tolerance for pipe based on manufacturing methods (e.g. seamless pipe and pipe made from plate).

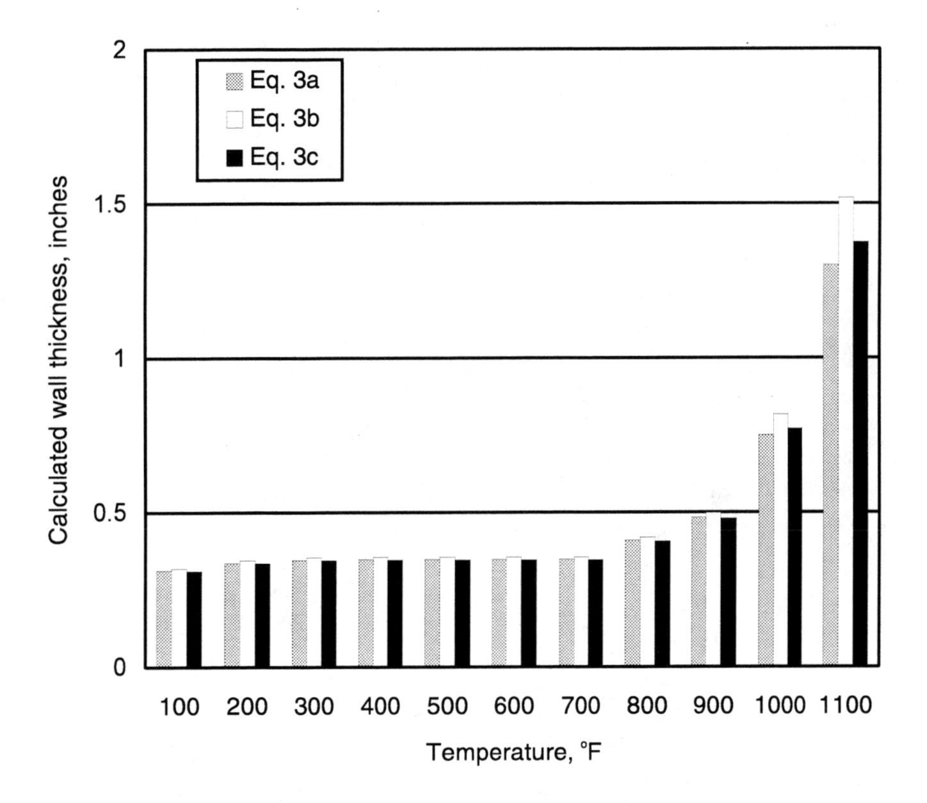

Figure 2.5 ASME B31.3 wall thickness OD equation values for ASTM A 335 P22, NPS 12, P = 1000 psig.

ASTM A 106 - 91 Standard Specification for Seamless Carbon Steel Pipe for High-Temperature Service

16. Permissible Variations in Weight and Dimensions

16.1 *Weight* - The weight of any length of pipe shall not vary more than 10% over and 3.5% under that specified. Unless otherwise agreed upon between the manufacturer and the purchaser, pipe in NPS 4 and smaller may be weighted in convenient lots; pipe larger than NPS 4 shall be weighted separately.

16.2 *Diameter* - Variations in outside diameter shall not exceed those specified in Table 5.

16.3 *Thickness* - The minimum wall thickness at any point shall not be more than 12.5% under the nominal wall thickness specified.

Note 7 - The minimum wall thicknesses on inspection of some of the available sizes are shown in Table X2.1

ASTM A 516/A 516M - 90 Standard Specification for Pressure Vessel Plates, Carbon Steel, for Moderate-and Lower-Temperature Service

2. Referenced Documents

2.1 ASTM Standards:
A 20/A 20M Specification for General Requirements for Steel Plates for Pressure Vessels

ASTM A 20/20M - 93 Standard Specification for General Requirements for Steel Plates for Pressure Vessels

1. Scope

1.1 This specification covers a group of common requirements which, unless otherwise specified in the individual material specification, shall apply to rolled steel plates for pressure vessels under each of the follow specifications issued by ASTM:

ANNEXES (Mandatory Information)

A1 Permissible Variations in Dimensions, etc. - inch-pound units

ASTM A20/20M-93 (Continued)

A1.1 Listed below are permissible variations in dimensions, and notch toughness information, express in inch-pound units of measurement.

Table A1.1 Permissible Variations in Thickness for Rectangular Plates

Note 1 - Permissible variation under specified thickness, 0.01 in.

Note 2 - Thickness to be measured at ⅜ in. from the longitudinal edge.

Note 3 - For thickness measured at any location other than that specified in Note 2, the permissible maximum over-tolerance shall be increased by 75%, rounded to the nearest 0.01 in.

Wall Thickness for External Pressure

The required minimum thickness of a pipe under external pressure, either seamless or with longitudinal butt joints, is determined by the procedure presented in Section VIII Division 1 Paragraph UG-28 for D_o/t greater than 10 [¶304.1.3]. This procedure is discussed below.

Nomenclature:

A = factor determined from 5-UGO-28.0, (Figure 2.6, top).

B = factor determined from the applicable material chart for maximum design metal temperature (Figure 2.6, bottom).

D_o = outside diameter of pipe (inches).

E = modulus of material elasticity at design temperature (psi) (from chart in B31.3 Appendix C6).

L = total length of the pipe, inches.

P = external design pressure, psi.

P_a = calculated value of maximum allowable external working pressure for the assumed value of "t," psi.

t = minimum required thickness of pipe, inches.

Procedure:

Assume a value for t and determine the ratios L/D_o and D_o/t.

1. Enter Fig. 5-UGO-28.0 - Appendix 5 (Figure 2.6 top) at the value of L/D_o. For values of L/D_o greater than 50, enter the chart at a value of L/D_o = 50. For values of L/D_o less than 0.05, enter the chart at a value of L/D_o = 0.05
2. Move vertically to the graphed line on the D_o/t graph for the value of L/D_o determined in (1). Note there are several D_o/t graphs in Section VIII, select the graph for the material under consideration. From the point of intersection, move horizontally to the left to determine the value of factor A.

3. Using the value of A from (2), enter the applicable material graph in Appendix 5 for the material under consideration. The material graph of this example is carbon steel to 300ºF. Again, several such graphs appear in Section VIII. Move vertically to an intersection with the material/temperature line for the design temperature. Interpolation may be made between lines for intermediate temperatures.
4. In cases where the value of A falls to the right of the end of the material/temperature line, assume an intersection with the horizontal projection of the upper end of the material/temperature line. For values of A falling to the left of the material/temperature line, see (7) below.
5. From the intersection obtained in (3), move horizontally to the left and read the value of factor B.
6. Using this value of B, calculate the value of the maximum allowable external working pressure P_a using the following formula:

$$P_a = \frac{4B}{3\left(\frac{Do}{t}\right)}$$

7. For values of A falling to the left of the applicable material/temperature line, the value of P_a can be calculated using the following formula:

$$P_a = \frac{2AE}{3\left(\frac{Do}{t}\right)}$$

8. Compare the calculated value of P_a obtained in (6) or (7) with P. If P_a is smaller than P, select a larger value for t and repeat the design procedure until a value of P_a is obtained that is equal to or greater than P.

Example: Find the required wall thickness for a long 10.75 in. OD carbon steel pipe operating with a 350 psig external pressure at 300°F.

1. Let t = 0.365 in., L/D_o = 50, then D_o/t = 29.45
2. A = 0.00122 from Fig. 5-UCS-28.0 (Figure 2.6, top)
3. B = 11600 from Fig. 5-UCS-28.1 (Figure 2.6, bottom)
4. Pa = 4 x 11600/(3 x 29.45) = 525 psig.

The 0.365 in. wall thickness is adequate to contain the 350 psig external pressure.

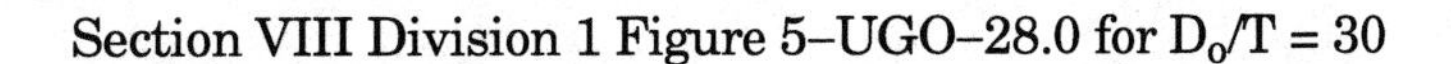

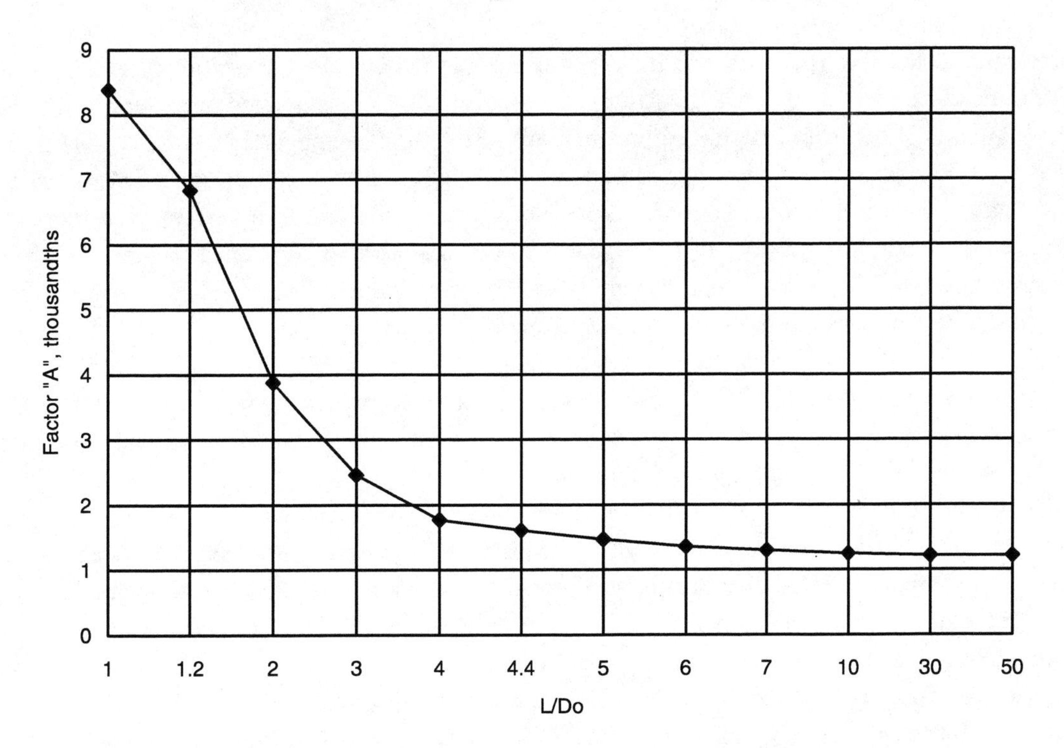

Section VIII Division 1 figure 5-UCS-28.1 Carbon steel to 300°F, E=29E6 psi

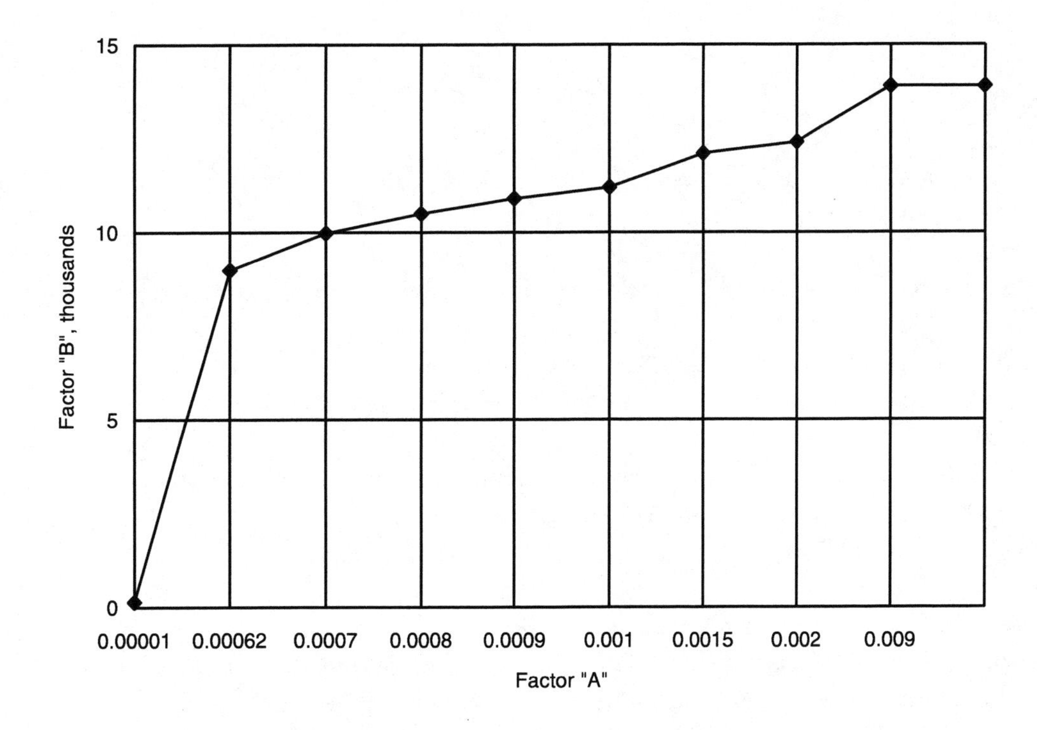

Figure 2.6 Factor "A" and factor "B".

Component Design

The preceding discussion reviewed the B31.3 design guidance for pressure piping. The required wall thickness calculation example above illustrates this in that the Code specified only a minimum wall and provided a means of determining that minimum wall. For the most part, this same type of guidance is provided for other pressure containing components as well. However, this does not mean that the designer has to perform a pressure design analysis of every component in a system. For most standard components, the Code has conducted a "prequalification" test that relieves the designer of the task of determining the components pressure design adequacy. In doing so, the Code has established three types of classifications for piping components as follows:

1. *Listed rated components* [¶302.2.1] are those for which the pressure-temperature rating has been established and, when used within these specified bounds, are considered safe with no additional need to determine the pressure design adequacy.
2. *Listed unrated components* [¶302.2.2] include components such as in-line fittings as elbows, reducers, and tees whose pressure rating is based on the pressure strength of matching seamless pipe. The component must be made of a material having at least the same allowable stress as the pipe and must have a pressure retention strength of at least 87.5% of the wall thickness of the matching pipe.
3. *Unlisted components* [¶302.2.3] may be used in a piping system provided the designer is satisfied with the chemical and physical properties of the material used in the component and with the quality control and method of manufacture. The pressure design of an unlisted component must then be proven (by calculations) to be in accordance with the pressure design in ¶304 or by ¶304.7.2 of B31.3.

Elbows and Bends

The Code pressure design requirements for elbows and bends that are not manufactured in accordance with listed standards are:

1. Bends can be made by either hot or cold processes, provided the material is suitable for the bending process [¶332.1].
2. The finished surface shall be free of cracks and substantially free from buckling.
3. The wall thickness after bending shall not be less than the minimum wall, considering corrosion-erosion and mill under-run tolerance.
4. Bend flattening, the difference between maximum and minimum diameters at any cross section shall not exceed 8% of nominal outside diameter for internal pressure and 3% for external pressure [¶332.2.1].

The major concern with respect to an elbow in external pressure service is the potential for a structural collapse of the elbow. Elbows in external pressure service more than 3% out-of-round may be well on the way to structural collapse. Elbows out-of-round as much as 8% in internal pressure service are permitted because the pressure will straighten the out-of-roundness to values less than 8%. Figure 2.7

shows the degree to which an elbow wall might be reduced on the extrados by the manufacturing process.

Miter bends have pressure limitation, as calculated by equations (4a), (4b), or (4c) of ¶304.2.3 of B31.3, which could derate a piping system. A miter is defined as an angle off-set greater than 3 degrees. See nomenclature for miter bends, Figure 2.8.

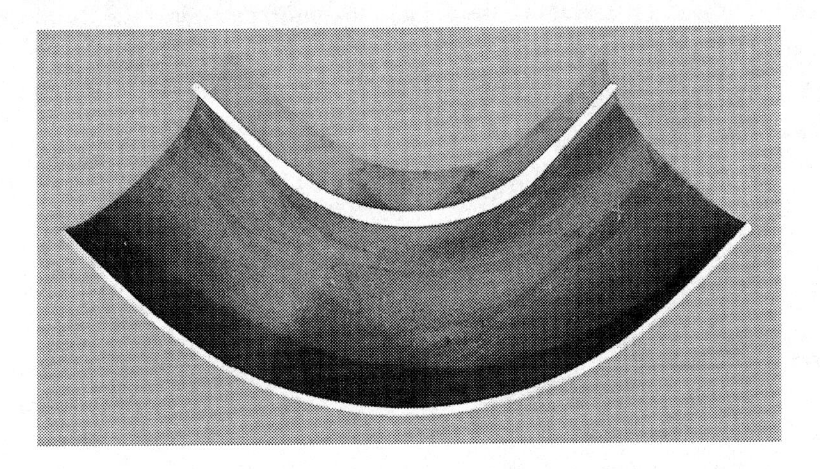

Figure 2.7 Elbow wall thickness thinning on extrados, thicknening on intrados.

Multiple miters, whose miter cut angle is less than 22.5 degrees are limited to a pressure that will generate hoop stresses not to exceed 50% of the yield strength of the material at temperature. This is done by restricting the maximum pressure to the lesser value as calculated by equation (4a) or (4b) in the Code.

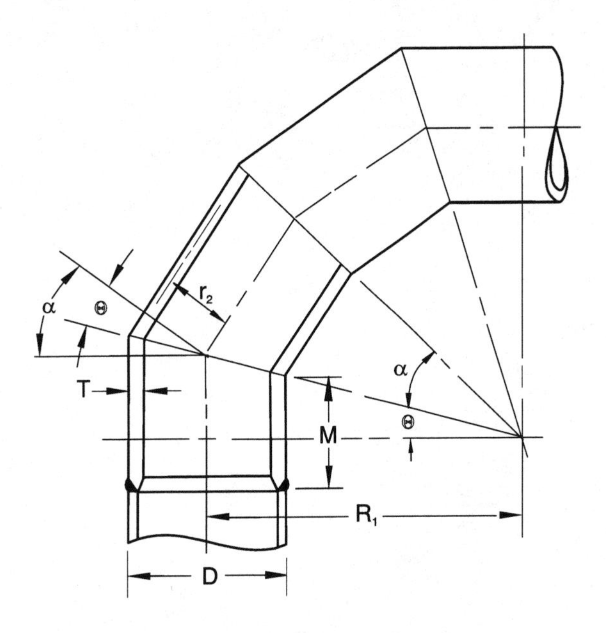

Figure 2.8 Nomenclature for miter bends

Single miters or miters whose bend angle is greater than 22.5 degrees are limited to hoop stresses of 20% of the material yield at temperature by equation (4c).

Designers who wish to use miters in a system but do not wish to pay this pressure penalty can simply increase the wall thickness of the miter, thus reducing the hoop stress to values less than the Code limit. Although this technique seems straight forward, it is not always clear where the miter starts.

The Code provides a method to determine how far the miter extends into the straight pipe. This distance is defined as "M" [¶304.2.3(c)], where

$M =$ the larger of $2.5 \times \sqrt{r_2 \times T}$ or

$\tan\theta \times (R1 - r_2)$ as shown in Fig. 2.8.

The "T" used in this equations is $\bar{T}$ less mill tolerance.

An example of miter bend maximum allowable internal pressure calculations per ¶304.2.3 follows:

Calculate the maximum allowable internal pressure for a 36 in. (0.375 in. nominal wall) miter bend constructed of A 515 Gr. 60 plate material, Temperature = 500°F, c = 0.10 in., E = 1.0 (fully radiographed), $R_1 = 1.5 \times 36 = 54$ in., $r_2 = 0.5\,(D_o - \bar{T})$, plate mill under-run tolerance = 0.01 in.

A. For $\theta = 22.5°$, use equation 4(a):

$$P_m = \frac{SE\,(T-c)}{r_2} \times \frac{T-c}{(T-c)+0.643\tan\theta\sqrt{(r_2(T-c))}}$$

at S = 17300 psi, E = 1.0, T = 0.375 - 0.010 = 0.365, $r_2 = 0.5(36 - 0.375) = 17.8125$

then $$P_m = \frac{17300 \times 0.265}{17.8125} \times \frac{0.265}{0.265+0.643\tan(22.5)\sqrt{(17.8125 \times 0.265)}}$$

$P_m = 80$ psig

For $\theta = 22.5°$, use equation (4b):

$$P_m = \frac{SE\,(T-c)}{r_2} \times \frac{R_1 - r_2}{R_1 - 0.5\,r_2}$$

$$P_m = \frac{17300 \times 0.265}{17.8125} \times \frac{54 - 17.8125}{54 - 0.5 \times 17.8125}$$

$P_m = 206$ psig

For the multiple miter elbow, the maximum allowable internal pressure is 80 psig.

B. For $\theta = 30°$ ($\theta > 22.5°$), $R_1 = 36$ in., use equation (4c):

$$P_m = \frac{SE\,(T - c)}{r_2} \times \frac{T - c}{(T - c) + 1.25 \tan\theta \sqrt{(r_2(T - c))}}$$

$$P_m = \frac{17300 \times 0.265}{17.8125} \times \frac{0.265}{0.265 + 1.25 \times 0.577 \sqrt{(17.8125 \times 0.265)}}$$

$P_m = 37$ psig

The maximum allowable internal pressure for this piping system is 37 psig. For compliance with B31.3 ¶304.2.3(d), the value of R_1 used in the above equations shall not be less than:

$$R_1 = \frac{A}{\tan\theta} + \frac{D}{2}$$, where the value of "A" is taken from the following:

(T-c), in.	A
≤ 0.5	1.0
$0.5 < (T\text{-}c) < 0.88$	2(T-c)
≥ 0.88	[2 (T-c)/3] + 1.17

The minimum R_1 for the above example then is as follows: at $(T\text{-}c) \leq 0.5$, $A = 1$, then $R_1 = 20.4$ in. for $\theta = 22.5°$ and the minimum R_1 for $\theta = 30°$ is 19.7 in.

Branch Connections

Piping system branch connections may be made by any one of several methods. These methods include tees, pad reinforced (see Figure 2.9) or unreinforced intersections, crosses, integrally reinforced weld-on or weld-in contoured insert fittings, or extrusions [¶304.3.1].

The Code philosophy for pressure design of intersections is concerned with the available pressure reinforcement offered by the geometry of the intersection. The process of making an intersection weakens the run pipe by the opening that must be made in it. Unless the wall thickness of the run pipe is sufficiently in excess of that required to sustain pressure at an intersection not manufactured in accordance with a listed standard, it may be necessary to provide added material as reinforcement. This reinforcement material is added local to the intersection, that is integral to the run and branch pipes.

Figure 2.9 Pad reinforced branch connection.

The amount of required pressure reinforcement is determined by performing area replacement calculations [¶304.3.3] using the design conditions established for the intersection. Area replacement calculations are not required for intersections using listed-rated or listed-unrated tee intersections provided the intersection is used within the pressure-temperature bounds stated in the listing standard. Area replacement calculations are not required for unlisted tee intersections, provided the tee component meets at least one of the requirements of ¶304.7.2, which are:

1. duplicating a successful operating system,
2. experimental stress analysis,
3. proof test.
4. detailed stress analysis (finite element)

Also, pressure reinforcement is not required for intersections using couplings whose size does not exceed NPS 2 nor one-fourth the nominal diameter of the run.

The B31.3 procedures for replacement calculations are valid for the conditions:

1. The center line of the branch pipe must intersect the center line of the run.
2. The angle of the intersection, β, (see Fig. 304.3.3 in B31.3) must be between 45° and 90°.
3. The run pipe diameter to thickness ratio, (D_h/T_h), is less than 100.
4. The branch to run diameter ratio (D_b/D_h) is not greater than 1.0.
5. For $D_h/T_h \geq 100$, the branch diameter can be no larger than one-half the run diameter.

Intersections that do require reinforcement calculations (those that are not qualified by ¶304.7.2 or by being listed), are qualified by summing all the integral metal around the intersection, (within a prescribed boundary), (reinforcement area) that is beyond that required to contain pressure and comparing that sum to the metal area removed to make the intersection. The required replacement area must at least equal that area removed to make the intersection. Consider the following two examples.

The first example is for a 90° unreinforced fabricated tee; the second is for a pad reinforced fabricated tee of the same pipe sizes and design conditions as the unreinforced fabricated tee (see Figure 2.10).

Find the area of replacement metal required in the unreinforced fabricated tee for the conditions:

Run pipe: NPS 8; schedule 40; ASTM A 53 Gr. B ERW
Branch pipe: NPS 4; schedule 40; ASTM A 53 Gr. B SMLS
P = 600 psig; T = 400°F; c = 0.10

At the time of the piping system design, it will not be known whether or not the intersection will fall on the longitudinal weld seam of the run. Therefore, the quality factor E is used to conservatively determine the minimum wall thickness of the run pipe. Had it been known that the intersection would not fall on the longitudinal weld seam, the E factor would not be required for the area replacement procedure, even though the run is ERW. [¶304.3.3]

Example A - area replacement calculation:
8.625 in. OD; $\bar{T}$ = 0.322 in. x 4.500 in. OD; $\bar{T}$ = 0.237 in., unreinforced fabricated tee.

I. Nomenclature. (Reference Fig. 304.3.3)

T = 400°F; P = 600 psig; c = 0.10 in.; β = 90°
D_h = 8.625; $\bar{T}_h$ = 0.322; Header Material: A 53 Gr. B ERW; E = 0.85
D_b = 4.500; $\bar{T}_b$ = 0.237; Branch Material: A 53 Gr. B SMLS; E = 1.0
Material SE; Header: 17,000 psi; Branch: 20,000 psi

T_h = 0.312 in. T_b = 0.207 in. ($\bar{T}$ - mill tolerance)

$d_1 = D_b - 2\,(T_b - c) = 4.5 - 2\,(0.207 - 0.10) = 4.286$ in. ← B31.3 DOES NOT GIVE THIS EQUATION (90° ONLY)

d_2 = the greater of d_1 or $(Tb - c) + (Th - c) + d_1/2$

d_2 = 4.286 in.

L_4 = the lesser of $2.5\,(T_h - c)$ or $2.5\,(T_b - c) + T_r$

$L_4 = 2.5\,(0.207 - 0.10) + 0 = 0.267$ in.

The pressure design thickness for the header and branch pipes, calculated using equation (3a):

$t = (P \times D)/2\,(SE + P \times Y)$; t_h = 0.150 in., t_b = 0.067 in.

II. Required Area

$A_1 = (t_h \times d_1) \times (2 - \sin\beta) = 0.643$ in.^{2}

III. Area Contributing to Reinforcement

$A_2 = (2 \times d_2 - d_1) \times (T_h - t_h - c) = 0.266 \text{ in.}^2$

$A_3 = 2 \times L_4 (T_b - t_b - c) = 0.021 \text{ in.}^2$

A_4 = (area of additional metal, including weld metal, within the reinforcing zone, t_c = 0.166 in. [¶328.5.4]) = 0.055 in.^{2}

$A_5 = A_2 + A_3 + A_4 = 0.342 \text{ in.}^2$

IV. Percent area replaced = $(A_5/A_1) \times 100$ = 53%

Example A intersection is not suitable for pressure design. Considering that the percent replaced area is only 53%, a reinforcing pad must be added to the intersection and area replacement calculations are repeated as follows in Example B. Had the replacement percent in example A been very near the 100% required, more weld metal could have possibly been added to attain the required 100%.

The retest with the pad in Example B yields 254% replaced area. The Code requirements for pressure design of the intersection are therefore satisfied.

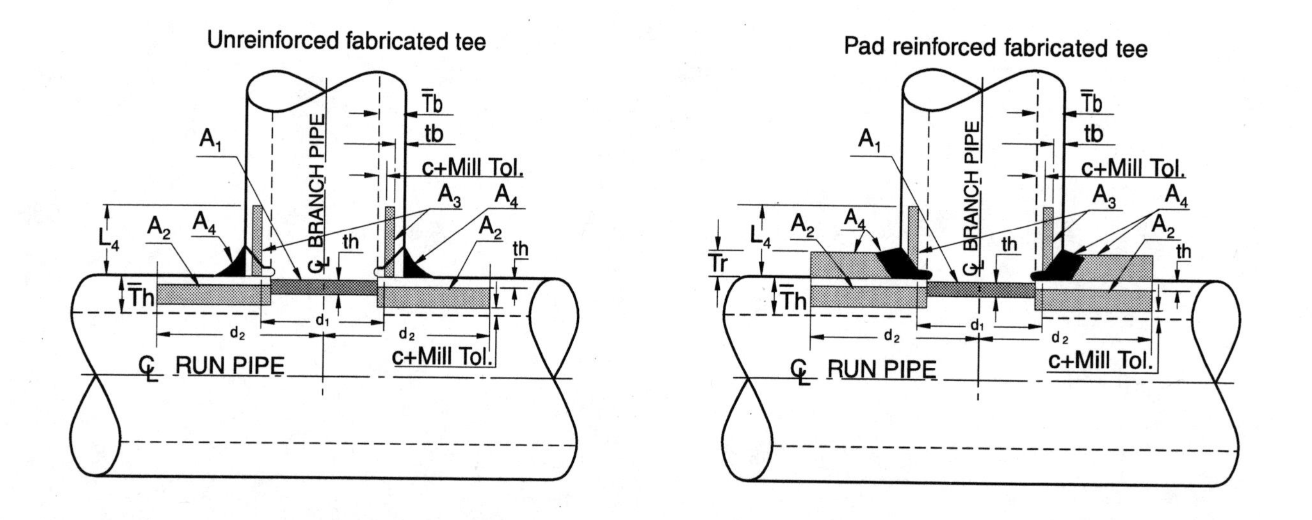

Figure 2.10 Branch connections.

Example B – Intersection:
8.625 in. OD; $\bar{T}$ = 0.322 in. x 4.500 in. OD; $\bar{T}$ = 0.237 in., pad reinforced intersection;
T_r = 0.322 - 0.01 = 0.312" (nominal wall less mill tolerance).

I. Nomenclature. (Reference Fig. 304.3.3)

T = 400°F; P = 600 psig; c = 0.10; T_r = 0.322
D_h = 8.625; $\bar{T}_h$ = 0.322; Header Material: A 53 Gr. B ERW; E = 0.85
D_b = 4.500; $\bar{T}_b$ = 0.237; Branch Material: A 53 Gr. B SMLS; E = 1.0
Material SE; Header: 17,000 psi; Branch: 20,000 psi

$T_h = 0.312$ $\quad$ $T_b = 0.207$ $\quad$ ($\bar{T}$ - mill tolerance)

$d_1 = D_b - 2\,(T_b - c) = 4.5 - 2\,(0.207 - 0.10) = 4.286$ in.

d_2 = the greater of d_1 or $(T_b - c) + (T_h - c) + d_1/2$

$d_2 = 4.286$ in.

L_4 = the lesser of $2.5\,(T_h - c)$ or $2.5\,(T_b - c) + T_r$

$L_4 = 2.5\,(0.312 - 0.10) = 0.530$ in.

The pressure design thickness for the header and branch pipes, calculated using equation (3a):

$t = (P \times D)/2\,(SE + P \times Y)$; $t_h = 0.150$ in., $t_b = 0.067$ in.

II. Required Area

$A_1 = (t_h \times d_1) \times (2 - \sin\beta) = 0.643$ in.^{2}

III. Area Contributing to Reinforcement

$A_2 = (2 \times d_2 - d_1) \times (T_h - t_h - c) = 0.266$ in.^{2}

$A_3 = 2 \times L_4\,(T_b - t_b - c) = 0.042$ in.^{2}

A_4 = (area of PAD, $T_r\,(2D_2 - D_b)$ = 1.270 in.^{2}, and weld metal, ($2t_c^2$ = 0.055 in.^{2}) within the reinforcing zone, t_c = 0.166 in.) = 1.325 in.^{2}

$A_5 = A_2 + A_3 + A_4 = 1.633$ in.^{2}

IV. Percent Area Replaced = $(A_5/A_1) \times 100$ = 254%

Extruded Outlet Header

The B31.3 ¶304.3.4 rules for the design of extruded outlet branch intersections are similar to the area replacement procedures of the preceding section. These rules are illustrated by the following example for an NPS 12 x 8 Standard extruded outlet intersection.

Conduct an area replacement calculation test on a 12.75 in. OD (0.375 in. nominal wall) x 8.625 in. OD (0.322 in. nominal wall) extruded outlet in accordance with ¶304.3.4. Assume the extruded geometry is the same as shown in Fig. 304.3.4 (c) of the Code. Refer to ¶304.3.4 (c) for nomenclature.

Design conditions: T = 500°F; P = 800 psig; c = 0.06 in.
Pipe material: ASTM A 106 Gr. B SMLS; S_h = 18,900 psi

Header	Branch
D_h = 12.75 in.	D_b = 8.625 in.
$\overline{T}_h$ = 0.375 in.	$\overline{T}_b$ = 0.322 in.
T_h = 0.328 in.*	T_b = 0.282 in.*
d_h = 12.094 in.	d_b = 8.061 in.

*Nominal wall less mill tolerance.

The die used for manufacture produced r_x = 0.75 in. and T_x = 0.813 in.

Applying the Code limits for r_x:

Minimum = the lesser of 0.05 D_b or 1.50 in.
Maximum = 0.1 D_b + 0.50

For this extrusion, 0.05 x 8.625 = 0.431 in. (minimum) and 0.1 x 8.625 + 0.50 = 1.363 in. (maximum). The r_x of 0.75 in. is within the Code limits.

Next, calculate the pressure design wall thickness for the header and branch pipes using Eq. (3d), the inside diameter wall thickness equation:

$$t = \frac{P\,(d + 2c)}{2(SE - P(1 - Y))}$$

where

Y = 0.4 and E = 1.0.
Header t_h = 800 (12.094 + 2 x 0.06)/[2 (18,900 - 800 (1 - 0.4))] = 0.265 in.
Branch t_b = 800 (8.061 + 2 x 0.06)/[2 (18,900 - 800 (1 - 0.4))] = 0.178 in.
(Note: Eq. (3a) produced t_h = 0.265 in., t_b = 0.179 in.)

Now calculate the required area, A_1.

$A_1 = Kt_hd_x$;
$d_x = d_b + 2c$ = 8.181 in.

To find K, where K is determined as follows:

1) for $D_b/D_h > 0.60$, K = 1.00
2) for $0.60 \geq D_b/D_h > 0.15$, $K = 0.60 + ⅔ (D_b/D_h)$

$D_b/D_h = 8.625/12.75 = 0.67$, K = 1.0

$A_1 = 1.0 \times 0.265 \times 8.181 = 2.168 \text{ in.}^2$

The excess metal in the header is A_2.

$A_2 = d_x (Th - t_h) = 8.181 (0.328 - 0.265-) = 0.515 \text{ in.}^2$

The excess metal in the branch pipe is A_3.

$A_3 = 2L_5 (T_b - t_b)$; $L_5 = 0.7 \sqrt{D_b T_x}$

where $T_x = T'_x - 0.06 = 0.813 - 0.06 = 0.753$ in., $(T'_x = T_x + c)$

then $L_5 = 0.7 \sqrt{8.625 \times 0.753} = 1.784$ in.

and $A_3 = 2 \times 1.784 (0.282 - 0.178) = 0.371 \text{ in.}^2$

The excess area in the extruded lip within r_x is A_4.

$A_4 = 2r_x (T_x - T_b) = 2 \times 0.75 (0.753 - 0.282) = 0.706 \text{ in.}^2$

% replaced area = $((A_2 + A_3 + A_4)/A_1)$ 100 = 73%.

This extrusion fails the area replacement test. Next manufacture the extrusion using dies to produce a geometry as shown in Fig. 304.3.4 (d) where the following dimensions are changed, as listed below. The sub-subscript '$_2$' is added to signify second test.

$d_{x_2} = d_x - 0.5 = 8.181 - 0.5 = 7.681$ in. (corrosion included)

$T_{x_2} = T'_x + 0.25 - 0.06 = 0.813 + 0.25 - 0.06 = 1.003$ in.

$r_{x_2} = 1.00$

Recalculate the area as follows:

$A_{1_2} = 1.0 \times 0.265 \times 7.681 = 2.035 \text{ in.}^2$

$A_{2_2} = 7.681 (0.328 - 0.265) = 0.484 \text{ in.}^2$

$L_{5_2} = 0.7 \sqrt{8.625 \times 1.003} = 2.058$ in.

$A_{3_2} = 2 \times 2.058 (0.282 - 0.178) = 0.428 \text{ in.}^2$

$A_{4_2} = 2 \times 1 (1.003 - 0.282) = 1.442 \text{ in.}^2$

% replaced area = $100 (A_{2_2} + A_{3_2} + A_{4_2})/A_{1_2}$ = 115%

This geometry is acceptable for pressure design.

Block Pattern Fittings

Block pattern pipe fittings that are normally found in high pressure systems (see Figure 2.11) can be designed by a pressure area procedure which employs a totally different methodology from that used by B31.3 for unreinforced fabricated tees, pads, or extrusions previously discussed in the Branch Connections section. In the pressure area procedure, the component is graphically designed (drawn) and the pressure area calculations are performed as illustrated in Figure 2.12. This procedure is taken from *Design of Piping Systems* by the M. W. Kellogg Company and is also found in ASME Section III, Subsection NB for design of wye pattern valves.

Figure 2.11 Block pattern wye.

B31.3 offers no guidance on the design of wye fittings by area replacement procedures. The tee and lateral can be designed using B31.3 procedures for pressure design of intersections because these procedures are valid for branch to run angles ranging from 90° to 45°. The Kellogg procedures are not B31.3 procedures; however, they are the procedures commonly used in the high pressure fitting manufacturing industry.

PRESSURE AREA TECHNIQUE

An example of the procedure for designing a wye is as follows (see Figure 2.12):

1. Prepare a scale drawing of the fitting using all known dimensions: the inside diameter, angle of intersection, pipe outside diameter, and wall thickness at the point of weld of all three pipes.
2. Estimate a "t_1" and "t_2" dimension. t_1 starting dimension could be about twice the wall thickness of pipe "D_1"; t_2 starting dimension could be about three times the wall of pipe "D_2".
3. Calculate lengths of pressure reinforcement zones using t_1 and t_2 in the equations.
4. Draw the fitting outside wall using scaled dimensions.

5. Calculate areas indicated and perform the pressure area stress test using equations "S_A" and "S_B". S_A and S_B are S_h, the basic allowable stress of the fitting material at temperature, taken from Table A-1 of B31.3.

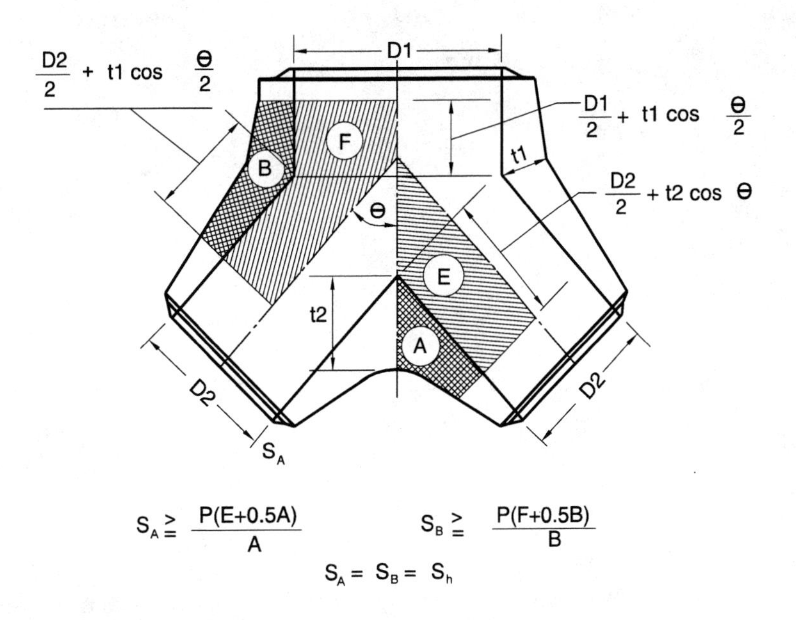

Figure 2.12 Illustration for block pattern wye design.

Two or three trial iterations may be required to obtain the necessary fitting wall thicknesses t_1 and t_2 to satisfy the S_A and S_B pressure area requirement and to obtain a suitable outside fitting surface to facilitate machining (manufacturing).

Closures

The Code approach for designing closures [¶304.4] is the same as for determining the wall thickness for pipe. The wall thickness selected for closures, considering pressure, corrosion-erosion, and mill under run tolerances, would be the next available plate or forging thickness. The pressure design wall thickness, "t", for closures from ASME Section VIII, Division 1 of the BPV code are presented below in Table 2.3. B31.3 does not provide equations for closures, the designer is referred to ASME Section VIII. These equations are presented for the purposed of illustrating their similarity to the pipe wall thickness equations discussed earlier. Terms not defined earlier are defined below in the table.

Other conditions may be applicable in the use of these equations. Refer to the BPV code before use.

Nomenclature:

L = inside spherical or crown radius, in.

α = one-half the apex angle in conical heads and sections, deg.

Table 2.3 Pressure Design Thickness

Type of Closure	Concave to Pressure	Convex to pressure
Ellipsoidal 2:1 Head	$\frac{PD}{(2SE-0.2P)}$	the larger of 1) concave "t", with P = 1.67Po, or 2) "t" from UG-33(d)*
Torispherical Head for 6% Knuckle Radius	$\frac{0.885PL}{(SE-0.1P)}$	the larger of 1) concave "t", with P = 1.67Po, or 2) "t" from UG-33(e)*
Hemispherical Head for t<0.365L and P<0.665SE	$\frac{PL}{(2SE-0.2P)}$	see UG-33(c)*
Conical Head (without transition knuckle)	$\frac{PD}{(2\cos(\alpha)(SE-0.6P))}$	see UG-33(f)*
* ASME Section VIII Division 1.		

Flanges

Generally, flanges used in refinery service are manufactured in accordance with the listed standard, ANSI/ASME B16.5. Flanges made to this standard and used within the pressure-temperature limits specified therein, are suitable for service without any further pressure design analysis.

The pressure-temperature limits are established for each flange pressure class and material. Examples of a few of the limits are shown in Table 2.4.

Table 2.4 Pressure-Temperature Limits

Pressure Class	Material	Maximum Temperature, °F	Maximum Pressure, psig
150	A105	100	285
300	A105	100	740
600	A105	100	1480
150	F316	100	275
300	F316	100	720
600	F316	100	1440

The equation for determining the pressure-temperature rating, taken from B16.5 is:

$$P(t) = \frac{P(r)S(1)}{8750}$$

where

P(t) = rated working pressure, psig, for the specified material at temperature.

P(r) = pressure rating class index expressed in psi. With the exception of class 150, P(r) is equal to the pressure class of the flange, i.e. P(r) = 300 for class 300 flanges. For class 150 flanges, P(r) = 115.

For temperatures below the creep range, S(1) = the lesser of:

1. 60% of the specified minimum yield strength at 100°F.
2. 60% of the yield strength at temperature.

For design temperatures in the creep range, S(1) value limits are listed in Annex D of the B16.5 Standard. However, refinery flanges will typically be based on the two conditions listed above for design temperatures less than 500 to 600°F.

The constant 8750 is approximately the average allowable stress of typical flange material at the highest rating temperature.

This flange rating equation can be used to establish the pressure/temperature ratings for a particular piping system. A typical application follows:

What is the required flange rating for a system with the following design conditions?

T = 400°F; P = 650 psig; Flange material: ASTM A 182 Gr. F11 CL 2

Solution: The yield strength at temperature S(y) (from ASME Section II Part D) is:

S(y) = 33,700 psi,

then S(1) = 20,220 psi.

(Note: A close approximation of the yield strength at temperature can be determined by dividing S_h of the flange material from Table A-1 of B31.3 by ⅔. In this example,

S_h = 22,500 psi

then S(y) = 22,500/0.66 = 33,783 psi.

This technique of determining the yield strength at temperature is valid only at temperatures below the creep range.

Rearranging the P(t) equation to solve for P(r):

$$P(r) = 8750 \times \frac{P(t)}{S(1)}$$

$$P(r) = 8750 \times \frac{650}{20{,}220}$$

P(r) = 281 class

Therefore, a class 300 flange would be selected for this service.

On occasion, flanges are subjected to high external piping loads (forces and moments) that may cause excessive leakage. In such situations, typical concerns are: How much external load is too much? How are flanges that are subjected to both internal pressure and external loads to be evaluated?

B31.3 refers the designer to Appendix 2 of the ASME Section VIII Division 1 Code for the design of flanges [¶304.5.1]. The Appendix 2 procedure considers only internal pressure to determine the adequacy of flanges. It does not consider external loads.

One very conservative procedure for considering external loads on flanges is to calculate the equivalent pressure from the external loads and combine this equivalent pressure (P_e) with the design pressure (P) of the piping system. Then perform the Appendix 2 analysis using $P_l = P_e + P$.

The equation to determine the equivalent pressure, "P_e"as presented by ASME Section III, Subsection NB is

$$P_e = \frac{16M}{\pi G^3}$$

where

M = overturning moments or torsional moments considered separately, in-lb
G = diameter at location of gasket load reaction, in. (G is a calculable number using the Appendix 2 equations).

An example of an application of the equivalent pressure calculation:

A flange has an external bending moment of 5,000 ft-lb resulting from thermal displacement in the pipe. What is the equivalent pressure for a flange with a G = 12 in.?

Solution: Convert the moment to in.-lb.

5,000 ft-lb = 60,000 in.-lb

then $$P_e = 16 \times \frac{60{,}000}{3.14 \times 12^3}$$

$$P_e = 177 \text{ psig}$$

The next step would be to add this P_e to the design pressure and perform the ASME Section VIII Division 1 - Appendix 2 analysis using the sum. Again, this is a very conservative approach. An example of the ASME Section VIII analysis follows.

ASME Section VIII Division 1 - 1992

Flange description: NPS 10 RF WN 150# flange
Flange location: low steam *Item no: 1*
Design pressure = 170 psig; temperature = 500°F
External moment = 24,000 in.-lb; equivalent pressure = 71.5 psig

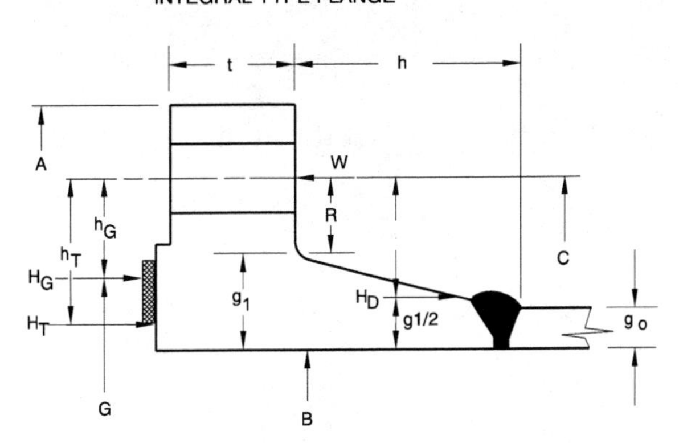

Figure 2.13 Flange dimensions.

A 105 flange material allowable stress:		*A 193-B7 bolt material allowable stress:*	
Ambient	*Design temperature*	*Ambient*	*Design temperature*
23,300 psi	*19,400 psi*	*25,000 psi*	*25,000 psi*

Gasket material: stainless steel
Gasket factors: m = 3; y = 10,000; b = 0.272; G = 11.956

Flange Dimensions (all dimensions in inches)

A = 16 *B = 10.02* *C = 14.25*
t = 1.125 *h = 2.329* *g1 = 0.99* *go = 0.365*
Bolt nom. dia. = 0.875 *number bolts = 12* *Fillet radius = 0.188*

Bolt load: operating = 41,898 lb *gasket seating = 114,002 lb*
Flange moments: operating = 60,965 in.-lb *gasket seating = 130,761 in.-lb*
Bolt areas: required = 4.084 in² *actual = 5.036 in.²*
Flange factors: L = 0.93 *K = 1.596* *ho = 1.912* *V = 0.081*
Modulus: Ec = 2.95E+07 *Eh = 2.73E+07*
Flange rigidity: operating = 0.132 *gasket seating = 0.265*

Stress Report

	Operating Stresses		***Gasket Seating Stresses***	
	Calculated	***Allowable***	***Calculated***	***Allowable***
SH	*6,674*	*29,100*	*14,314.7*	*34,950*
SR	*7,864.7*	*19,400*	*16,868.4*	*23,300*
ST	*2,769.9*	*19,400*	*5,941.1*	*23,300*
(SH+SR)/2	*7,269.3*	*19,400*	*15,591.5*	*23,300*
(SH+ST)/2	*4,722*	*19,400*	*10,127.9*	*23,300*

Flange weight = 54.1 lb

Flanges designed by the ASME Section VIII Division 1 Appendix 2 method above may not be sufficiently rigid with respect to flange ring rotation. As bolt loads are applied to seat the gasket, flange ring rotation may occur which may then affect the gasket seating surface and gasket efficiency in controlling leakage. ASME Section VIII - Division 1 recognizes this potential for flange leakage and has included in the non-mandatory Appendix S-2, a procedure for testing the flange rigidity for two types of flanges - integral hub type and loose ring type, with and without hubs.

The limit of permissible ring rotation during gasket seating or normal operation is:

Flange Type	**Permissible Ring Rotation**
Integral Hub	0.3°
Loose Ring	0.2°

To determine flange rigidity, calculate the flange rigidity index "J" where J must be ≤ 1.0. The terms of these equations are defined in Appendix S-2.

J for integral hub flanges is:

$$J = \frac{52.14MoV}{0.3LEgo^2ho}$$

J for loose ring flanges with hubs:

$$J = \frac{52.14MoVl}{0.2LEgo^2ho}$$

J for loose ring flanges without hubs:

$$J = \frac{109.4Mo}{0.2Et^3Ln(K)}$$

This check is to be conducted at both gasket seating and operating conditions.

An example of the application of these equations follows, using the preceding Appendix 2 example. The values of the J equation terms are presented in Table 2.5 below. This example is for an integral hub flange.

Table 2.5 Values for Rigidity Factor J

	Gasket Seating	**Operating**
Mo	130761 in-lb	60965 in-lb
V	0.081	0.081
L	0.93	0.93
go	0.365	0.365
ho	1.912	1.912
E	$29.5x10^6$	$27.3x10^6$
J	0.265	0.132

This flange has sufficient rigidity to limit J to ≤ 1.0.

Another method of analyzing flanges subjected to bending moments is presented in Appendix IV of the ASME B31 Mechanical Design Committee, Model Chapter II document. This method estimates the moment to produce leakage of a flange joint with a gasket having no self-sealing characteristics.

The equation is:

$$M_L = \left(\frac{C}{4}\right)\left(S_b A_b - P A_p\right)$$

where

M_L = moment to produce leakage, in-lb
C = bolt circle, in.
S_b = bolt stress, psi
A_b = total area of flange bolts at root of thread, in.^{2}
P = internal pressure, psi
A_p = area to outside of gasket contact, in.^{2}

As example of an application of this technique, find the leakage moment for the following conditions:

Flange = 150; class ANSI B16.5 standard bore
Temperature = 100°F; pressure = 285 psig
Corrosion = 0; gasket seating stress on bolts = 30,000 psi
Gasket dimension from Flexitallic General Catalog.
The pipe bending stress resulting from the bending moment is not intensified. SIF = 1.0.

Table 2.6 M_L, Leakage Moments

	Bolt				M_L Leakage	Pipe Bending
NPS	**Circle**	**Root Dia.**	**No.**	**Gasket OD**	**Moment, in.-lb**	**Stress, psi**
2	4.75	0.507	4	3.375	25,728	45,861
3	6.00	0.507	4	4.750	28,749	16,676
4	7.50	0.507	8	5.875	76,324	23,777
6	9.50	0.620	8	8.250	135,835	15,981
8	11.75	0.620	8	10.375	141,996	8,447
10	14.25	0.731	12	12.500	413,440	13,827
12	17.00	0.731	12	14.750	434,929	9,930
14	18.75	0.838	12	16.000	661,784	12,416
16	21.25	0.838	16	18.250	1,009859	14,365
18	22.75	0.963	16	20.750	1,439,535	16,066
20	25.00	0.963	20	22.750	2,006,245	18,009
24	29.50	1.088	20	27.000	2,909,053	17,968

One conclusion that can be drawn from Table 2.6 is that with the selection of an appropriate safety factor (about 2 or 3), designers can generate a rule of thumb to analyze flanges. Dividing the bending stress by these factors of safety, a bending stress of about 6,000 psi will result, meaning that for bending stress less than 6,000 psi, no further analysis may be required for the flange.

The graph in Figure 2.14 can be used to improve the design of a flange where the longitudinal hub stress, S_h is too high and the f factor is >1.0. Recognizing that f = 1.0 is optimum design, for flange shapes which produce f >1.0, possibly the hub is too short. By increasing the hub length, the f factor will approach 1.0.

f is the ratio of the maximum axial bending stress at the juncture of the hub and ring. The axial bending is expressed in terms of S_h, longitudinal hub stress (Figure 2.14).

$$S_h = \frac{fM_o}{Lg_1^2B}$$

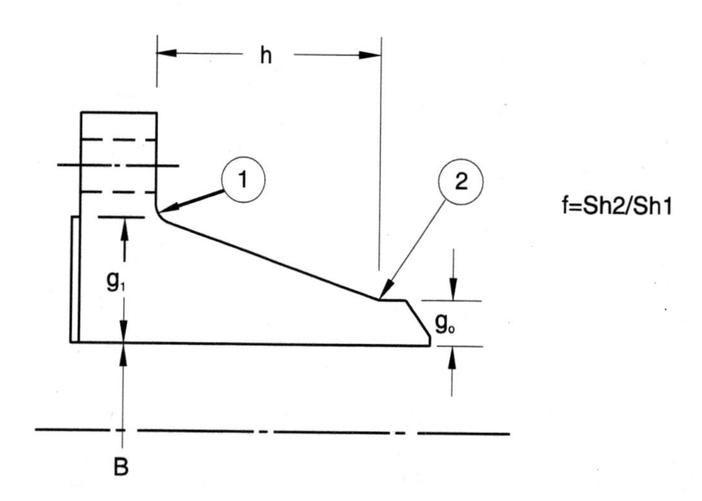

Blind Flanges

The thickness, t, of blind flanges for B31.3 service is determined in accordance with ASME Section VIII Division 1, UG-34 [¶304.5.2]. There, the equations for several styles of blind flanges are presented, each with its appropriate "corner factor," C.

The thickness of a bolted blind flange is determined by the UG-34 equation:

$$t(m) = t + c$$

where $$t = d\sqrt{\frac{CP}{SE} + 1.9\frac{Wh_G}{SEd^3}}$$

d = mean diameter of the gasket seating surface (inches)
C = 0.3 for bolted flange corner condition (see Fig UG-34 for Bolted Flanges)
P = pressure, psig
S = basic allowable stress at temperature (psi)
E = joint efficiency
W = total bolt load (lb)
h_G = bolt moment arm (bolt circle diameter - d)/2 (inches)

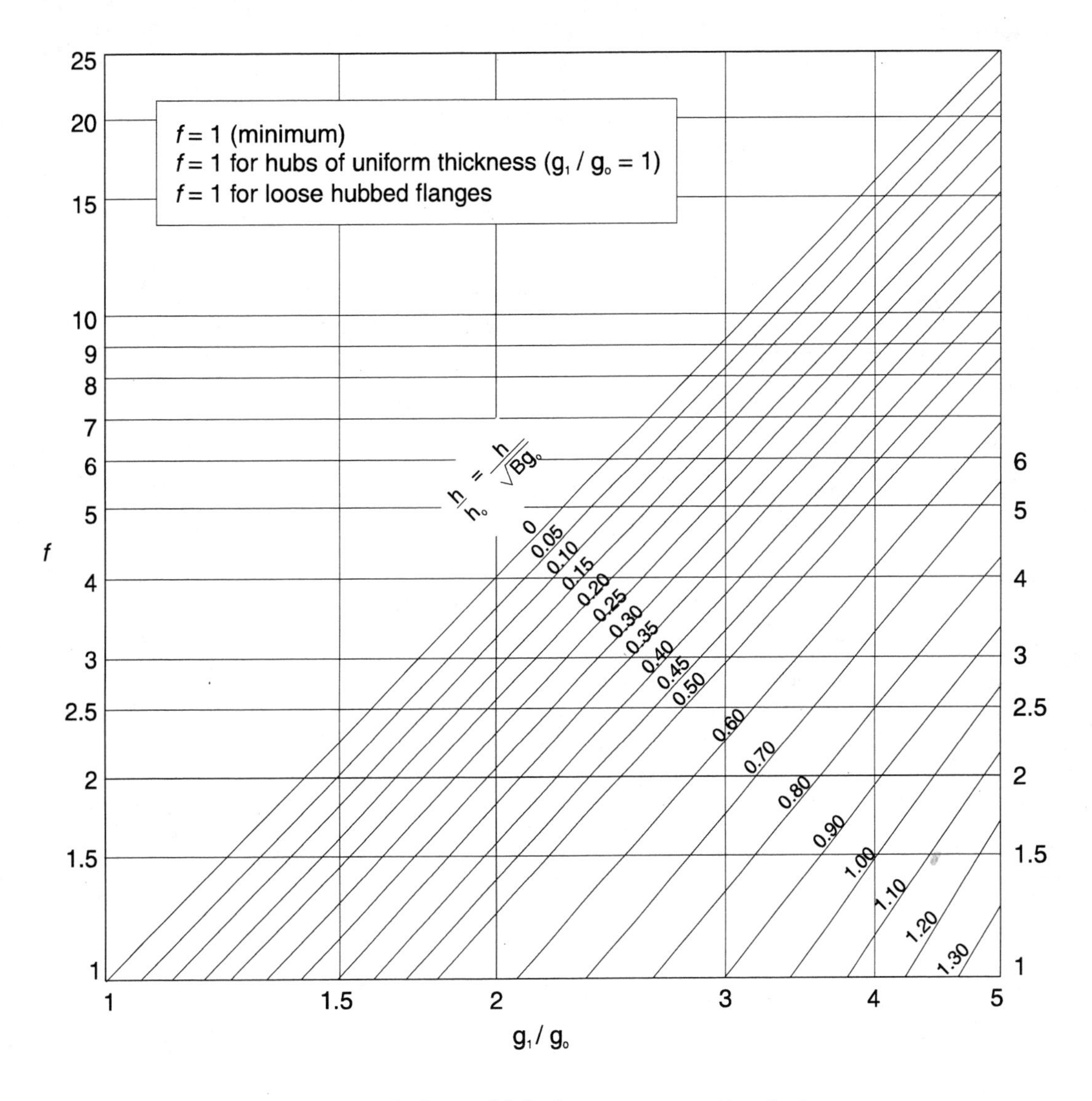

Figure 2.14 Values of *f*: hub stress correction factor.

As an example of the application of these equations, consider the following example:

You are asked to determine the pressure design adequacy of a seamless NPS 10 in. ASTM A 105 bolted blind flange with a measured thickness of 1.625 in. The flange dimensions and design conditions are:

T = 100°F; d = 11.385 in.; 8 - ¾ in. diameter bolts; P = 700 psig; C = 0.30; bolt circle = 14.25 in.
Corrosion allowance = 0.063 in.

The installation record from the job site stated that the bolts were tightened to 100 ft-lb torque using moly-cote lubricant, K = 0.129.

Does the blind flange have adequate thickness?

Solution:

S = 23,300 psi (S_h from Appendix A for ASTM A 105 material)
E = 1.0, seamless flange.
h_G = (14.25 - 11.385)/2 = 1.4325 inches
W, the total bolt load, can be calculated from the bolt torque equation:

$$\tau = KDF_p$$

where

D = Nominal bolt diameter, in.
K = Lubrication factor.
F_p = Bolt pull per bolt, lbs. or, 100 ft-lb = 1200 in.-lb = 0.129 x 0.75 x F_p
F_p = 12,403 lb per bolt,

then W = 8 bolts x 12,403 lb/bolt
W = 99,224 lb

Solve the expressions of the t equation:

$$\frac{CP}{SE} = \frac{0.3 \times 700}{23{,}300 \times 1} = 0.00901$$

$$\frac{1.9Wh_G}{SEd^3} = \frac{1.9 \times 99{,}224 \times 1.4325}{23{,}300 \times 1 \times 11.385^3} = 0.00785$$

Substitute these expressions and solve for t:

$$t = 11.385 \times \sqrt{0.00901 + 0.00785}$$

$$t = 1.478 \text{ in.}$$

finally t_m = 1.478 + 0.063 = 1.541 in.

The measured thickness of 1.625 in. is greater than the Section VIII required thickness of 1.541 in. The blind flange is therefore suitable for the service.

Blanks

Blanks are used in piping systems to isolate temporarily the flow to or from a section of the piping. They act as a blind flange in that they are exposed to the full longitudinal pressure force; the thickness of the blank must be adequate to contain this pressure force. B31.3 provides an equation [¶304.5.3] to evaluate the pressure design thickness, t_m, of blanks as:

$$t_m = t + c$$

$$t = d_g \sqrt{\frac{3P}{16\,SE}}$$

where

d_g = inside diameter of gasket for raised face flanges, or the gasket pitch diameter for ring joint and fully retained gasketed flanges (inches)

E = quality factor from Table A-1A or A-1B
P = design gage pressure (psig)
S = stress value for material from Table A-1 (psi)
c = corrosion/erosion, plus mill tolerance allowance (inches)

An example of the application of the blank thickness equation follows:

Determine the required thickness, t_m, of an ASTM A 285 Gr. B seamless material blank in a 10.75 in. pipe, ID = 10.02 in.

P = 600 psig; T = 400°F; S = 15,400 psi (from Table A-1)
c = 0.063 in., consider both sides of the blank exposed to the corrosive environment. Mill under run tolerance is 0.010 in.
E = 1.0, see Table A-1B (Although ASTM A 285 Gr. B is not listed in this table, it can be seen that all seamless materials have an E factor of 1.0.)
d_g = 10.52 in., assume gasket inside diameter is equal to the pipe ID plus 0.5 in.

then $$t = 10.52 \sqrt{\frac{3 \times 600}{16 \times 15{,}400}} = 0.899 \text{ in.}$$

$$c = 2 \times 0.063 + 0.010 = 0.136 \text{ in.}$$

$$t_m = 0.899 + 0.136 = 1.035 \text{ in.}$$

A plate thickness of 1.035 in. would be required for this blank.

Expansion Joints

Expansion joints are used in piping for any of the following reasons:

1. to reduce expansion stresses,
2. to reduce piping reactions on connecting equipment,
3. to reduce pressure drop in a system where insufficient flexibility exist, and
4. to isolate mechanical vibration.

Other factors which could influence the decision to use expansion joints could be inadequate space or economics (favoring an expansion joint over a conventional expansion loop or off-set piping arrangement).

The two styles of expansion joints most frequently found in chemical plant and refinery service are *bellows* and *slip-joint* styles [¶304.7.4], when used alone or in combination service (i.e. universal, hinged, or gimbal bellows joint) can adequately absorb thermal expansion or contraction displacements. Table 2.7 serves as a guide for expansion joint selection for a given set of thermal displacements. Tables 2.8 and 2.9 on the following pages provide more information on bellows type expansion joints[1].

Table 2.7 Guide For Expansion Joint Selection

Displacement	Bellows	Slip-joint
Axial	Yes	Yes
Lateral	Yes	No
Angular rotation	Yes	No
Torsion	No	Yes

[1] Edgar, D.L. and Paulin Jr., A.W, "Modeling Metallic Bellows Type Expansion Joints In A Computerized Pipe Flexibility Analysis", ASME Pub - Vol. 168, Book No. H00483 - 1989.

Table 2.8 Modeling Metallic Bellows Type Expansion Joints
In A Computerized Pipe Flexibility Analysis

Unrestrained no hardware	
Limit rod	
Single slotted hinge	
Universal slotted hinge	
Single tied 2 rods	
Universal tied 2 rods	
Single tied 3 or more rods	
Universal tied 3 or more rods	
Single hinge	
Single Gimbal	
Pressure balanced tee/elbow	

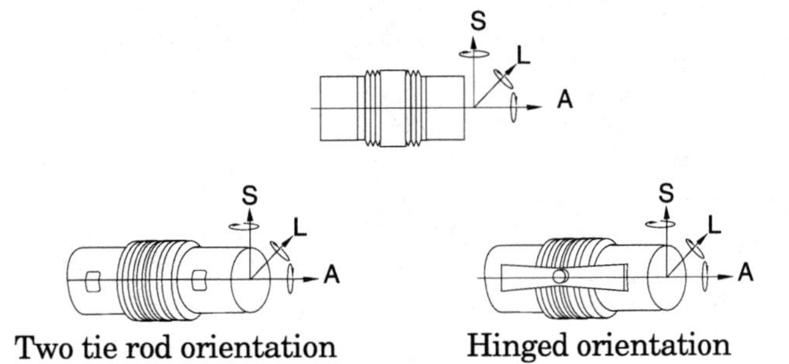

Two tie rod orientation Hinged orientation

A–axis – This is the longitudinal or "axial" expansion joint axis.

L–axis – This is the "lateral" axis, which is perpendicular to the A–axis, and in the plane of the restraining hardware, if any. If there is complete symmetry in the hardware, or if there is no hardware, then this axis may be arbitrarily selected.

S–axis – This is the "shear" axis of the assembly. It is perpendicular to the other two axes. The S–axis and L–axis are interchangeable for most types of expansion joints, the exceptions are two tie rods and hinged joints.

D_a - Axial deflection of the expansion joint,
D_l - Lateral translation of the expansion joint,
D_s - Shear translation of the expansion joint,
O_a - Rotation of the expansion joint about the longitudinal axis,
O_l - Rotation of the expansion joint about the L–axis,
O_s - Rotation about the S–axis,

An "L" in any column indicates that the value is subject to maximum limitations.

Table 2.9 Bellows Type Expansion Joint

Expansion Joint Style	**P_t at Anchor**	**Deflections**						**Stiffness Coefficient**						**Cost Fact.**	**Shell Temp**
		D_a	**D_s**	**D_l**	**O_a**	**O_s**	**O_l**	**S_a**	**S_s**	**S_l**	**SO_a**	**SO_s**	**SO_l**		
Single	Y	Y	Y	Y	Y	Y	Y	Y	Y	Y	Y	Y	Y	1	Hot
Two Limit Rod	Y	L	Y	L	Y	L	Y	L	Y	L	Y	L	Y	2	Hot
Single Tied 2 Rods	N	N	Y	Y	Y	N	Y	N	Y	Y	Y	N	Y	2-5	Cold
Single Slotted Hinge	Y	L	N	N	N	L	Y	L	N	N	N	L	Y	4	Hot
Singled Tied 3 Or More Rods	N	N	Y	Y	Y	N	N	N	Y	Y	Y	N	N	3-7	Cold
Single Hinge	N	N	N	N	N	N	Y	N	N	N	N	N	Y	6	Cold
Single Gimbal	N	N	N	N	N	Y	Y	N	N	N	N	Y	Y	8	Cold
Universal Tied 2 Rods	N	N	Y	Y	Y	N	Y	N	Y	Y	Y	N	Y	2-5	Cold
Universal Slotted Hinge	Y	L	Y	L	N	L	Y	L	Y	L	N	L	Y	6	Hot
Universal Tied 3 Or More Rods	N	N	Y	Y	Y	N	N	N	Y	Y	Y	N	N	3-7	Cold
Universal Hinge	N	N	Y	N	N	N	Y	N	Y	N	N	N	Y	12	Cold
Universal Gimbal	N	N	Y	Y	N	Y	Y	N	Y	Y	N	Y	Y	16	Cold
Pressure Balanced Tee Or Elbow	N	Y	Y	Y	Y	Y	Y	Y	Y	Y	Y	Y	Y	15-20	Cold
Hinge And Two Gimbal	N	Y	Y	Y	N	Y	Y	See Above for a Discussion of Individual Units						15-20	Cold

The support design of piping containing expansion joints requires special consideration for the anchors and guides used to control the thermal displacements and pressure forces. Main anchors used with unrestrained expansion joints are of particular importance and are the summation of three load functions:

a) sliding friction force of the pipe, F_f
b) spring force of the expansion joint, S_f and
c) pressure thrust force of the expansion joint, P_f

The sliding friction force, F_f, is equal to the weight of the sliding pipe times the friction factor coefficient. The spring force, S_f, is equal to the spring rate of the expansion joint times the displacement of the joint. These two forces are easily calculated. The pressure force, P_f, is often misunderstood and misused, therefore, a greater explanation of how to determine this force will be presented.

To understand this pressure force, consider a length of pipe with capped ends subjected to internal pressure (Figure 2.15a). The pressure force, P_f, on each capped end is:

$$P_f = \frac{\pi P D^2}{4}$$

where D = inside pipe diameter and P = design pressure.

The longitudinal pressure force is carried in tension in the pipe wall and the pipe is structurally stable. Now consider the same pipe with a band cut out of the center, then sealed with a flexible element, such as a bellows, which will not offer any resistance to the longitudinal pressure force (Figure 2.15b). As pressure is added to this bellowed pipe, the bellows, which would then be subjected to the longitudinal tensile force, will straighten out to a straight tube (Figure 2.15c). The bellows would no longer be capable of accommodating axial motion and may rupture because the wall thickness of the bellows is normally less than one-quarter that of the pipe. Hence, it is apparent that in order to maintain the bellows configuration, external main anchors must be added to carry the pressure force (Figure 2.15d). Such a system is called a *compression system*, since an external compressive force must be added to the piping system to prevent the bellows from elongating (see Figure 2.16 and Figure 2.17).

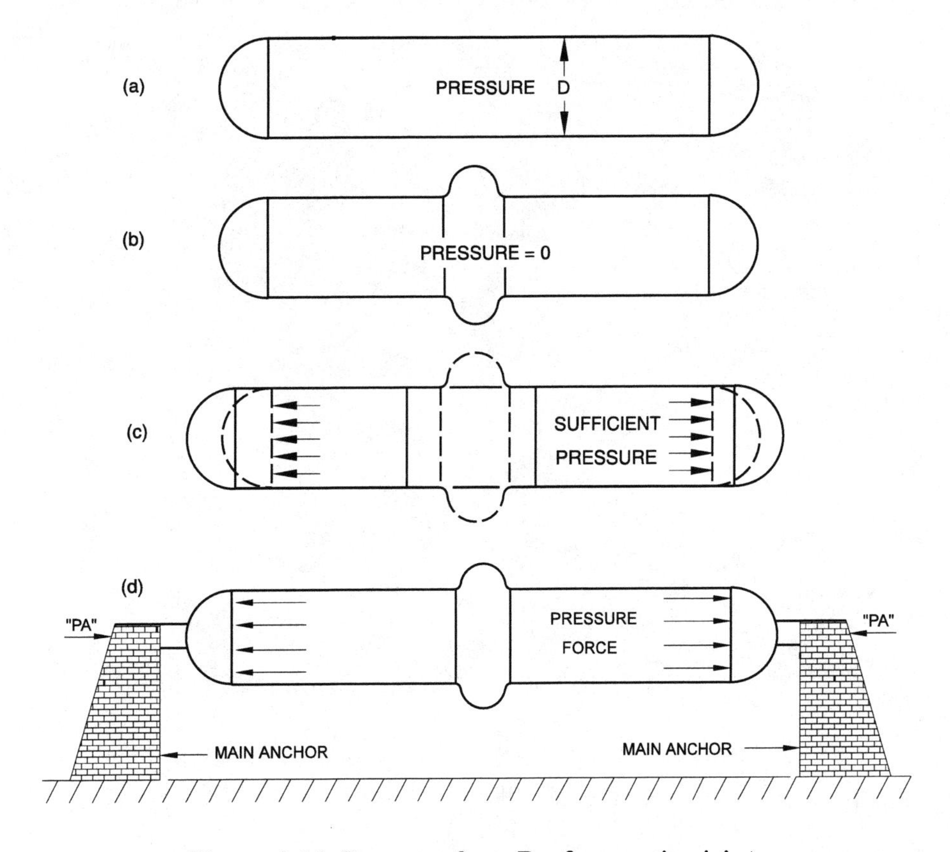

Figure 2.15 Pressure force P_f of expansion joint.

Figure 2.16 Bulldozer as main anchor.

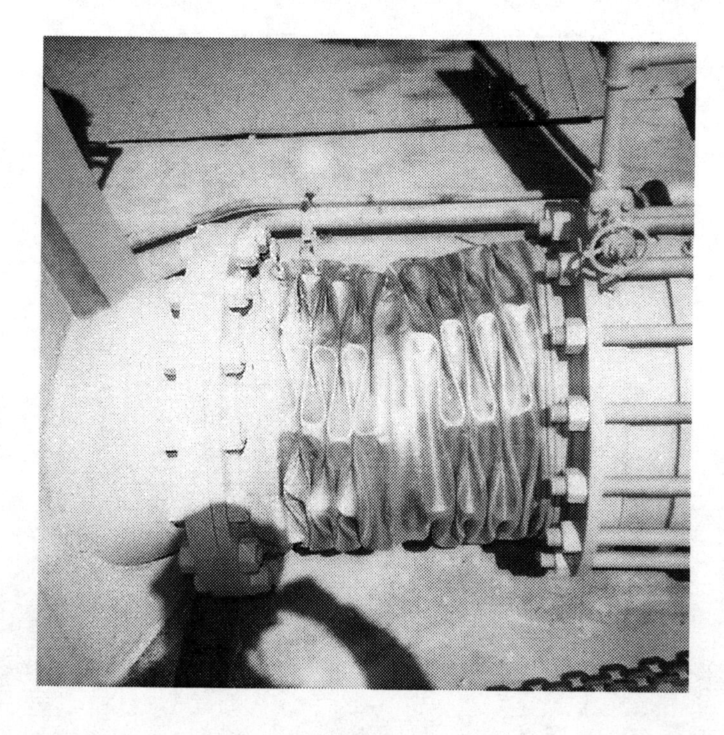

Figure 2.17 Bellows expansion joint after bulldozer pushed the pipe back to its original position.

The pressure force is determined by first calculating the force offered by the bellows inside diameter fluid area and adding this force to the bellows side wall thrust area (as detailed below).

The bellows inside diameter fluid area is:

$$\text{Bellows ID area} = \frac{\pi\,(ID)^2}{4}$$

The side wall thrust area can be visualized as internal pressure acts on the side wall of the bellows. The convolution would tend to spread out in the longitudinal direction. This force must be restrained by the main anchor as well. As an approximation, consider a 1 in. wide radial strip cut out of a corrugation and construct a free-body diagram of this strip when acted on by pressure (see Figure 2.18). For a fixed ended beam with uniform load,

$$R_{ID} = R_{OD} = Ph/2$$

Hence, one-half of the convolution side-wall load is carried in tension at the crest of the convolution and the other half is carried in compression at the inside diameter of the bellows.

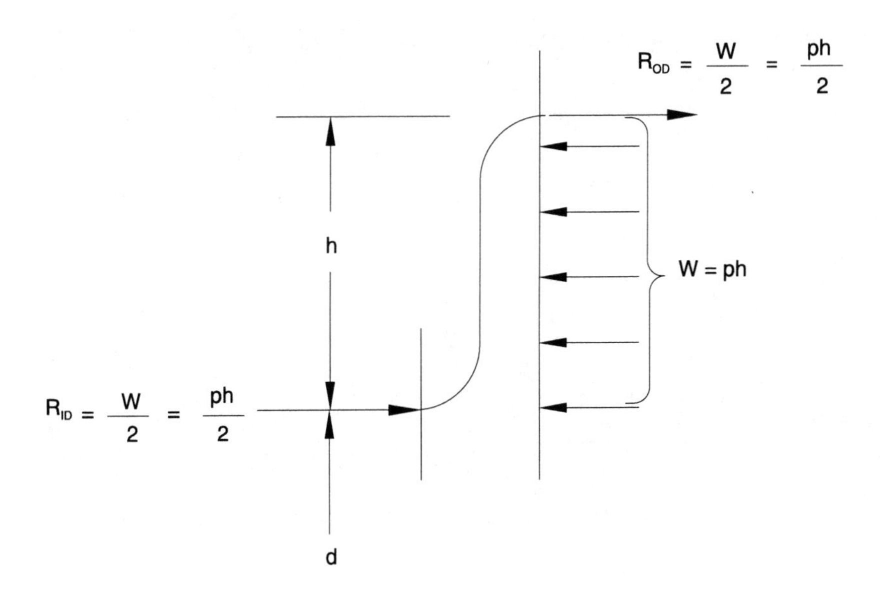

Figure 2.18 Free-body diagram.

The side wall effective area, SW_a, is

$$SW_a = \frac{\pi\left[\left(d + \frac{2h}{2}\right)^2 - d^2\right]}{4}$$

However, $d + h = D_m$, the mean diameter of the bellows.

Therefore, the side wall effective area = $\dfrac{\pi\left(D_m^2 - d^2\right)}{4}$

Then, the total effective pressure thrust area, A_e is;

$$A_e = \text{I.D. area} + \text{side wall area}$$

$$A_e = \frac{\pi\left(d^2 + D_m^2 - d^2\right)}{4}$$

$$A_e = \frac{\pi D_m^2}{4}$$

Finally, the total pressure force P_f is:

$$P_f = PA_e$$

Using the main anchor force procedure above, solve the following sample problem:

A 10.75 in. OD schedule 40 carbon steel piping system containing an unrestrained bellows expansion joint is designed to absorb axial expansion. The joint is placed midway between two main anchors that are 150 feet apart (Figure 2.19). The piping system is properly supported and guided. What is the force on the anchor, A_f?

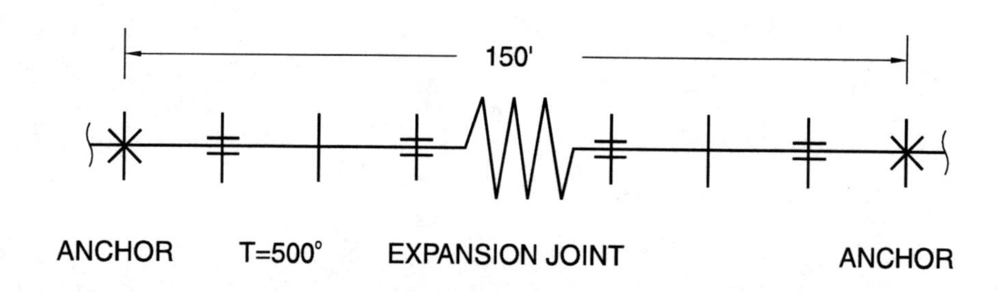

Figure 2.19 Bellows expansion joint centered in a piping length.

Design conditions:

T = 500°F; installed temperature = 70°F; bellows spring rate = 10,000 lb/in.
P = 200 psig; pipe contents = gas
Pipe + insulation weight = 49.43 lb/ft
Bellows ID = pipe ID = d
Bellows convolution OD = 14 in.
Friction coefficient at supports = 0.3

Solution: The anchor force A_f is the sum of friction force + bellows spring force + pressure force.

a) Friction force, F_f

Total friction force of the entire system is $F_f = 150 \times 49.43 \times 0.3 = 2{,}224.5$ lbf

b) Bellows spring force, S_f

The spring force is the total displacement across the 150 feet times the spring rate of the bellows. The total displacement is calculated as the expansion rate times the length of pipe expanding. The expansion rate, from Table C-1 of the B31.3, is:

$$e = 3.62 \text{ in/100 ft.}$$

Then the total expansion = 150 x 3.62/100 = 5.43 in.

Now $S_f = 5.43 \times 10{,}000 = 54{,}300$ lbf

c) Bellows pressure force, P_f

$$\text{The bellows effective pressure area, } A_e = \frac{\pi\left[10.02 + 0.5\,(14 - 10.02)\right]^2}{4} = 153.938 \text{ in.}^2$$

then $P_f = 153.938 \times 200 = 30{,}787$ lbf

Each anchor will see one-half of the total friction force, F_f, the total bellows spring force, S_f, and the total bellows pressure force, P_f. Therefore, the force on each anchor, A_f, will be:

$$A_f = 0.5 \times F_f + S_f + P_f$$

$$A_f = 0.5\,(2{,}224.5) + 54{,}300 + 30{,}787$$

$$A_f = 86{,}199 \text{ lbf}$$

Proper *guide spacing* is essential for the successful operation of expansion joints. The Expansion Joint Manufacturers Association, Inc. (EJMA) recommends a guide spacing from the expansion joint based on the pipe diameter, D, as follows.

- 1st guide at a distance of 4D from the joint,
- 2nd guide at a distance of 14D from 1st guide,
- All other guides at distances no greater than L_{max}.

L_{max} (feet) is calculated from the equation $L_{max} = 0.131\sqrt{\dfrac{EI}{(PA_e \pm fe_x)}}$

where

E = modulus of elasticity of the pipe material, psi
I = moment of inertia of pipe, in.^{4}
P = design pressure, psig
f = bellows initial spring rate per convolution, lb/in./conv.
e_x = axial stroke of bellows per convolution, in./conv.
A_e = effective pressure thrust area

Note: When bellows is compressed in operation, use (+)fe_x; when extended, use (-)fe_x.

Determine the guide spacing for the example problem above.

Assume the bellows had five (5) convolutions, each with a spring rate of S_f = 2,000 lb/in.,
$E = 27.9 \times 10^6$; I = 161 in.^{4}; (P = 200 psig; A_e = 153 938 in.^{2}; f = 2000)
e_x = 5.43/5 = 1.086 in. axial stroke per convolution.

Then $$L_{max} = 0.131\sqrt{\frac{27.9 \times 10^6 \times 161}{200 \times 153.938 + 2{,}000 \times 1.086}}$$

$$L_{max} = 48.4 \text{ ft}$$

The guide spacing for this problem would be:

1st guide at 4 x 10/12 = 3.33 ft
2nd guide at 14 x 10/12 = 11.66 ft
All remaining guides at 48.4 feet maximum spacing.

CHAPTER

3

FLEXIBILITY ANALYSIS OF PIPING SYSTEMS

The safety of a piping system subjected to a temperature change and resulting thermal displacement is determined by a *flexibility analysis* to insure against the following [¶319.1.1]:

1. Overstrain of piping components,
2. Overstrain of supporting structures,
3. Leakage at joints, and
4. Overstrain of connecting equipment, without material waste.

Required Analysis

Compliance with B31.3 Code flexibility analysis is a requirement of most petroleum and chemical plant piping installations. The Code places the burden of this analysis the designer [¶300 (2)] and holds the designer responsible to the owner for assuring that all the engineering design complies with the requirements of the Code.

The Code is clear as to which piping systems require an analysis; all systems require an analysis with the exception of the following: [¶319.4.1]

1. Those that are duplicates of successfully operating installations,
2. Those that can be judged adequate by comparison with previously analyzed systems, and
3. Systems of uniform size that have no more than two anchor points, no intermediate restraints, and fall within the limitation of the equation:

$$\frac{Dy}{(L - U)^2} \leq K_1$$

where

D = outside diameter of pipe, in. (mm)
y = resultant total displacement strains, in. (mm), to be absorbed by the piping system
L = developed length of piping between anchors, ft (m)
U = anchor distance, straight line between anchors, ft (m)
K_1 = 0.03 for U.S. customary units listed above (208.3 for SI units).

Using the above equation, is a flexibility analysis required for the following installation?

A two-anchor routing of a 8.625 in. outside diameter, schedule 40 carbon steel pipe is shown in Figure 3.1 below. The design temperature is 200°F, installed temperature = 70°F.
(e = 0.99 in./100 ft at 200°F, fromTable C-1).

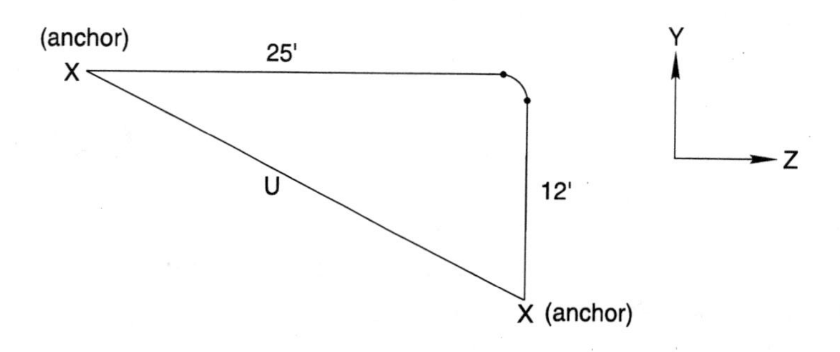

Figure 3.1 Two-anchor piping system.

D = 8.625 in.

$$y = \sqrt{\Delta Y^2 + \Delta Z^2}$$

$\Delta Y = 12 \times 0.99/100 = 0.1188$ in.

$\Delta Z = 25 \times 0.99/100 = 0.2475$ in.

$$y = \sqrt{(0.1188)^2 + (0.2475)^2} = 0.2745 \text{ in.}$$

L = 12 + 25 = 37 ft

$$U = \sqrt{(12)^2 + (25)^2} = 27.73 \text{ ft}$$

then $8.625 \times 0.2745/(37 - 27.72)^2 = 0.0275$, so

$Dy/(L - U)^2 \leq 0.03$

This two-anchor problem falls within the limits of the equation and does not require any further thermal fatigue analysis.

Although this simple equation is useful in determining the need for formal stress analysis, it does have limitations. No general proof can be offered to assure that the formula will yield accurate or conservative results. Users are advised to be cautious in applying it to abnormal configurations (such as unequal leg U-bends with L/U greater than 2.5 or near-saw-tooth configurations), to large diameter thin-wall pipe (stress intensification factors of the order of 5 or more), or to conditions where extraneous motions other than in the direction connecting the anchor points constitute a large proportion of the expansion duty.

Allowable Stress Range

B31.3 establishes maximum allowable stress limits that can be safely accommodated by a piping system before failure will commence for two separate stress loading conditions. These limits are for stress levels that can cause failure from a single loading, S_h, and those that can cause failure from repeated cyclic loadings, S_A.

The allowable stress range, S_A, [¶302.3.5 (d)] is the stress limit for those stresses that are repeated and cyclic in nature, or simply, it is the allowable stress to be compared to the calculated *displacement stress range,* "S_E" [¶319.4.4]. S_E (a secondary stress) will be discussed in the Displacement Stress Range section of this chapter.

The allowable stress range is presented in B31.3 by two equations:

Equation (1a):

$$S_A = f\,(1.25\,S_c + 0.25\,S_h)$$

S_A, by equation (1a), is a "system" allowable stress of the entire piping system of the same material and temperature.

and equation (1b):

$$S_A = f\,[1.25\,(S_c + S_h) - S_L]$$

S_A, by equation (1b), is a "component" allowable stress at temperature where S_L has been calculated for that component.

S_c and S_h are the basic allowable stresses for the cold and hot conditions as defined in the Defintion and Basis for Allowable Stress section in Chapter 1. Their values are found in B31.3 Appendix A Table A-1. (**Note**: For cryogenic or cold pipe service, S_c is taken at the operating temperature, S_h is taken at the installed temperature).

f is the *stress-range reduction factor* presented in B31.3 Table 302.3.5 or equation (1c).

S_L is the *longitudinal stresses* to be discussed later in the Sustained Load Stress section in this chapter.

An example of the application of the allowable stress range equation (1a) is as follows:

Calculate the S_A for a piping system constructed of ASTM A 106 Grade B pipe material used in 500°F service, and with a design life of 18,000 thermal cycles.

Solution: From B31.3 Table A-1 for ASTM A 106 Grade B

S_c = 20,000 psi, (at min. temp. to 100°F)

S_h = 18,900 psi, (at 500°F)

f = 0.8 (from B31.3 Table 302.3.5), then

$S_A = 0.8\,(1.25 \times 20{,}000 + 0.25 \times 18{,}900)$

S_A = 23,780 psi

This piping system can be expected to operate safely provided the displacement stress range, S_E, does not exceed S_A of 23,780 psi and the number of thermal cycles is less than 18,000. (The f factor although appropriate for the 18,000 cycles of this problem, is also suitable for 22,000 cycles as shown in Table 302.3.5.)

The allowable stress range equation (1b) can be used as a design basis in place of equation (1a) provided the longitudinal stresses due to sustained loads, S_L, have been calculated for each component and these longitudinal stresses are less than the hot allowable stress, S_h, ($S_L \leq S_h$).

As an example of equation (1b): Assume the S_L in the above (1a) example was calculated to be 8,000 psi (S_L = 8,000) at an elbow in a piping system. Using equation (1b), what is the new S_A for the elbow?

$S_A = 0.8\,[1.25\,(20{,}000 + 18{,}900) - 8{,}000]$

S_A = 32,500 psi.

The allowable stress range is increased by nearly 25% by including the unused allowable stress for sustained loads in this example. It is interesting to note that in this example, S_A is above the yield strength at temperature. (Yield = 18,900/0.66 = 28,636 psi). How can an allowable stress be higher than the yield strength of a material at temperature? The answer is that, this allowable stress is for a secondary stress which is self-limiting and will diminish in time through local yielding of the stressed components in the piping system, such as elbows or branch connections.

Note again, equation (1a) is a *system allowable stress*, the allowable stress of the entire piping system of the same material and temperature. Equation (1b) is a *component allowable stress*, the allowable stress of every single component in a piping system where S_L has been calculated.

On occasion, the thermal cycles a particular piping system will experience in the life of the plant may vary from one or more sustained operating states. The system may undergo several plant upgrades where thermal cycling at the new operating temperature may occur for several years before changing to still another operating temperature where thermal cycling would continue to occur. After two or three of these new thermal cycling operating states, it would appear that the stress-range reduction factor, f, would be difficult to determine. If this were the case, then the remaining life of the piping system would be impossible to predict, because the remaining life is very much dependent upon the prior number of thermal cycles. How does one determine the f factor in such a piping system?

The answer to this question is found in B31.3. The Code provides an equation to calculate the *equivalent full temperature cycles* [¶302.3.5] for such operating conditions. The equation is:

$$N = N_E + r_1^5 N_1 + r_2^5 N_2 + \ldots + r_n^5 N_n$$

for i = 1, 2, ...n

where

N = number of equivalent full temperature cycles
N_E = number of cycles of maximum computed displacement stress range S_E
N_i = number of cycles associated with displacement stress range S_i
$r_i = S_i/S_E$

As an example of the determination of the equivalent cycles, consider the following:

What are the equivalent cycles of a system that operates for:

Cycles	S_i (psi)
5,000	15,000
3,000	10,000
1,000	8,000
500	5,000

Assume the installed temperature is 70°F.

Then $N_E = 5,000$
$N_1 = 3,000$
$r_1 = 10,000/15,000 = 0.666$
$N_2 = 1,000$
$r_2 = 8,000/15,000 = 0.533$
$N_3 = 500$
$r_3 = 5,000/15,000 = 0.333$

Finally $N = 5,000 + (0.666)^5 \times (3,000) + (0.533)^5 \times (1,000) + (0.333)^5 \times (500)$
$N = 5,438$ equivalent thermal cycles.

Now the allowable stress range can be accurately calculated for this piping system using the equivalent cycles to determine the stress range reduction factor.

Returning to B31.3 Table 302.3.5 we find that the stress range reduction factor f ranges from a value of 1 for 7,000 cycles or less to a value of 0.3 for 2,000,000 cycles. For piping experiencing more than 7,000 cycles in the life of the piping system, f can be calculated by the equation:

$$f = 6.0N^{-0.2} \leq 1.0$$

An example of the use of this equation is:

What is f for a system with 53,000 life time cycles?

$$f = 6.0\ (53{,}000)^{-0.2} = 0.68$$

Designers using B31.3 Table 302.3.5 would have a tendency to use f = 0.6. Here we can see the allowable stress range, S_A would be 13% higher (0.68/0.6 x 100 = 1.13%) using the calculated f in place of the non-interpolated table value. See Figure 3.2 for the graphical representation of f.

This equation for calculating f, although entirely appropriate for this purpose, is not entirely accurate. Consider a piping system with 7,000 life time cycles. The above equation will yield an f value of 1.02 (f = 1.02), where the table value is 1.0 (f = 1.0), which is the correct value for 7,000 cycles. For the equation to yield the correct f value, the constant will have to be changed from 6.0 to 5.875. However, using a three decimal accuracy in calculating f will imply an accuracy in B31.3 philosophy of calculating stresses that is not present. Using the constant 6 will produce a reasonably close f factor, one that is consistent with the accuracy of the simplified approach of the B31.3 Code.

The reason for selecting 7,000 cycles as a starting point for the determination of f was to simplify the calculations of typical piping systems. This 7,000 cycles represents roughly one cycle per day for a period of twenty years. Very few petroleum refineries will ever see 7,000 cycles. However, some chemical plants, particularly batch operated plants, are more likely to see more than 7,000 cycles.

Displacement Stress Range

The displacement stress range, S_E, is the calculated range of (secondary) stress a piping system will experience when subjected to thermal expansion or contraction. The temperature range for this calculation is the total expansion range from minimum to maximum for hot operating systems and from maximum to minimum for cryogenic or cold pipe. Pressure and weight (primary) stresses are not considered in this evaluation.

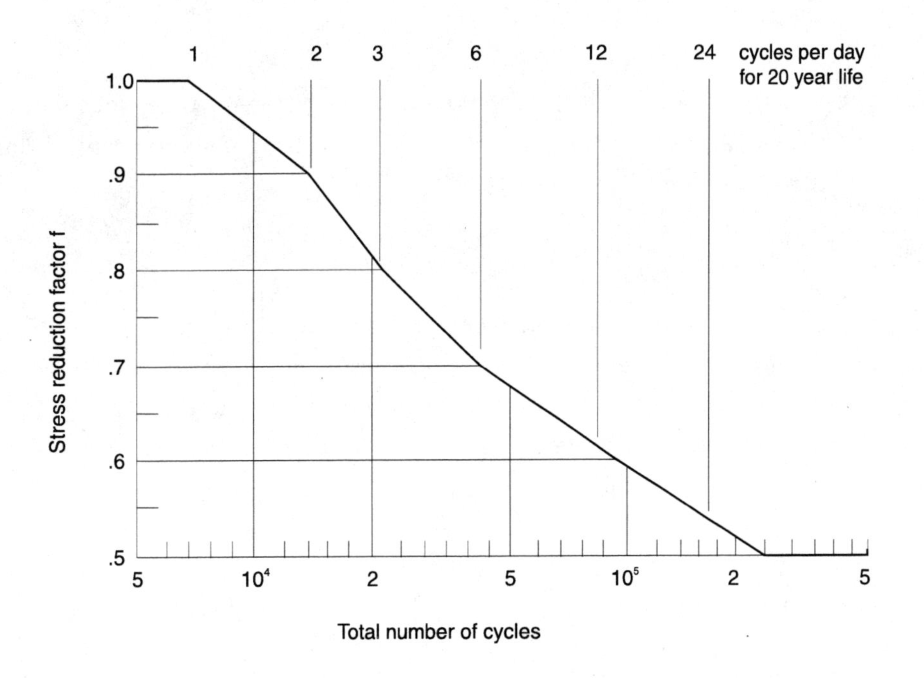

Figure 3.2 Plot of stress-reduction factor f.

The allowable stress range, S_A is the stress limit for comparison to the calculated displacement stress range, S_E. The B31.3 equation for the displacement stress range [¶319.4.4] is:

$$S_E = \sqrt{S_b^2 + 4S_t^2}$$

where

$$S_b = \text{resultant bending stress} = \frac{\sqrt{(i_i M_i)^2 + (i_o M_o)^2}}{Z}$$

$$S_t = \text{torsional stress} = \frac{M_t}{2Z}$$

M_t = torsional moment

Z = section modulus of pipe, in.^{3} = $\pi r_2^2 \overline{T}$, (the approximate equation, r_2 = mean pipe radius)

i_i = in-plane stress intensification factor from Appendix D

i_o = out-plane stress intensification factor from Appendix D

M_i = in-plane bending moment, in.-lb

M_o = out-plane bending moment, in.-lb

S_E is calculated using the pipe nominal wall thickness dimensions. Corrosion, erosion and mill tolerance are not subtracted from the pipe nominal wall thickness for S_E calculations.

Bending Stress

The bending stress component, S_b, [¶319.4.4(b)] of the displacement stress range, equation S_E, is the resultant of in-plane and out-plane bending moments due to thermal expansion or contraction.

Torsional Stress

The torsional stress component, S_t, of the displacement stress range is calculated by dividing the torsional moment by twice the section modulus of the pipe experiencing the torsion. Torsional stress at branch intersections caused by torsional moments in the branch pipe, and the section modulus are calculated using the branch pipe outside diameter and "$\bar{T}_b$". The torsional stress in the header pipe is calculated using the outside diameter and wall thickness, "$\bar{T}_h$" of the header pipe.

Consider the following example to illustrate the calculation of the bending stress, the torsional stress, and the displacement stress ranges.

A 10.75 in. schedule 40 LR elbow (nominal wall thickness = 0.365 in.) is subjected to forces caused by thermal expansion. The resulting bending and torsional moments from these forces are calculated as:

M_i = 100,000 in-lb $\quad$ M_o = 25,000 in-lb $\quad$ M_t = 9,000 in-lb

The section modulus, Z = 30.9 in.^{3}

What is the displacement stress range?

Solution: The SIF for the 10.75 in. LR elbow is calculated from B31.3 Appendix D as follows.

$i_i = 0.9/h^{2/3}$ for the in-plane SIF
$i_o = 0.75/h^{2/3}$ for the out-plane SIF

where

$$h = \frac{\bar{T}R_1}{r_2^2}$$

$\bar{T}$ = nominal wall thickness of the header pipe
R_1 = bend radius of elbow, in.
r_2 = mean radius of header pipe

then $\bar{T} = 0.365$; $R_1 = 1.5 \times 10 = 15$
$r_2 = 0.5\ (10.75 - 0.365) = 5.1925$
$h = 0.365 \times 15/(5.1925)^2 = 0.203$
$i_i = 0.9/0.203^{2/3} = 2.60$
$i_o = 0.75/0.203^{2/3} = 2.168$

The bending stress, S_b, can now be calculated.

$$S_b = \frac{\sqrt{(2.6 \text{ x } 100{,}000)^2 + (2.168 \text{ x } 25{,}000)^2}}{30.9}$$

$$S_b = 8{,}595 \text{ psi}$$

The torsional stress, $S_t = 9{,}000/(2 \text{ x } 30.9) = 145$ psi.

Finally $S_E = \sqrt{(8{,}595)^2 + 4 \text{ x } (145)^2}$

$$S_E = 8{,}600 \text{ psi}$$

For most piping systems in the moderate temperature range, the allowable stress range, S_A, using equation (1a), would be in the 28,000 to 30,000 psi range. In this example, the elbow is operating at a S_E of about 30% of the allowable stress range, S_A.

The equation for S_E is based on the maximum shear (TRESCA) failure theory and for convenient comparison with Code allowable stress range S_A, S_E represents two times the maximum shear stress due to expansion loading. S_E can be derived using the Mohr's Circle (Figure 3.3).

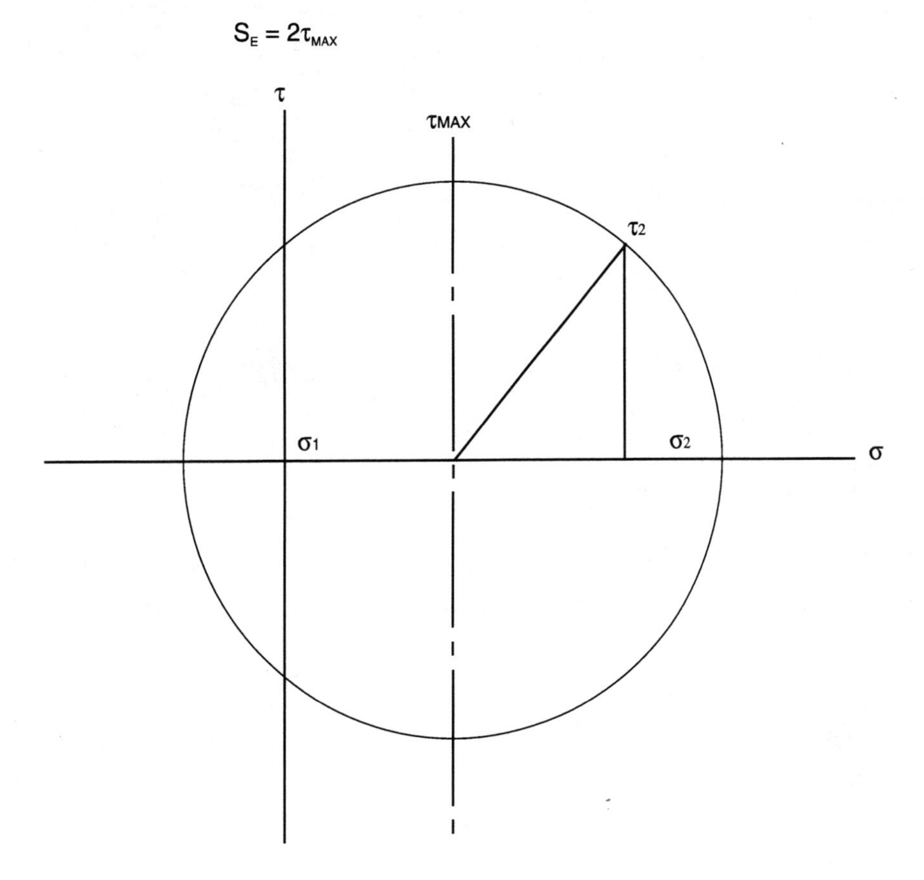

Figure 3.3 Mohr's Circle.

The stresses located on the principal stress axis, σ, are bending stresses. The stresses located on τ axis are torsional stresses.

σ_1 = least principal (bending) stress
σ_2 = maximum principal (bending) stress
τ = torsion (shear) at stress condition 2 (at σ_2)
$(\sigma_2 - \sigma_1)/2 = S_b/2$ (location of maximum shear in Mohr's Circle)

τ_{max} is located at a distance equal to the hypotenuse of the right triangle with $S_b/2$ and τ_2 as the triangle legs.

$$\tau_{max} = \sqrt{(S_b/2)^2 + (\tau_2)^2}$$

$$S_E = 2\tau_{max} = \sqrt{S_b^2 + 4\tau^2}$$

In the calculation of S_E, the section modulus, Z, used in the above example for calculating S_b is only valid for full size components. It is not valid for reducing outlet branch connections with forces and moments applied through the branch pipe. In the analysis of reducing outlet intersections, B31.3 instructs the designer to use Z_e, the *effective section modulus* instead of Z [¶319.4.4(c)].

The reason for using Z_e instead of Z is to adjust the calculated stresses to a more realistic stress magnitude. Tests have proven that for reducing intersections, using Z_e will result in calculated stresses closer to the actual measured stresses than would be calculated using Z.

The effective section modulus equation is:

$$Z_e = \pi r_2^2 T_s$$

where

T_S = the effective branch wall thickness, the lesser of $\overline{T}_h$ or $(i_i)(\overline{T}_b)$.
r_2 = mean radius of branch pipe = $0.5\,(D_b - \overline{T}_b)$

An example of the application of the effective section modulus follows:

Calculate S_E for a 12.75 in. outside diameter (0.375 in. nominal wall) x 6.625 in. outside diameter (0.28 in. nominal wall) standard B. W. Pipet[1] that is subjected to the same thermal forces and moments through the branch pipe as the elbow example.

M_i = 100,000 in.-lb
M_o = 25,000 in.-lb
M_t = 9,000 in.-lb

---

[1]Manufactured by WFI International, Houston, Texas.

Solution: The SIF is calculated from B31.3 Appendix D for the branch welded-on fitting (integrally reinforced) as:

$$i_i = i_o = 0.9/h^{2/3}$$

where

$h = 3.3\overline{T}/r_2$ and, for this example, $\overline{T}$ = 0.375 in., r_2 is the mean radius of run pipe.

$r_2 = 0.5\ (12.75 - 0.375) = 6.\ 1875$

then $i_i = i_o = 2.628$

Next, for Z_e calculation, determine the lesser of

$\overline{T}_h = 0.375$ or

$i_i\overline{T}_b = 2.628 \times 0.280 = 0.738$ (the branch pipe wall thickness is 0.28 in.)

Using the lesser, $\overline{T}_h$, Z_e can be calculated using the mean radius of the branch pipe, r_2.

$r_2 = 0.5\ (6.625 - 0.280) = 3.1725$

The effective section modulus can now be calculated

$$Z_e = \pi r_2^2\ \overline{T}_h = 11.857 \text{ in.}^3$$

The bending stress, S_b, is calculated as:

$$S_b = \frac{\sqrt{(2.628 \times 100{,}000)^2 + (2.628 \times 25{,}000)^2}}{11.857}$$

$$S_b = 22{,}846 \text{ psi}$$

The torsional stress from the branch is calculated using the branch section modulus determined by the equation, $Z_b = \pi r_2^2\ \overline{T}_b$ in.^{3} The term r_2 is the mean radius of the branch pipe.

$r_2 = 0.5\ (6.625 - 0.28) = 3.173$

$Z_b = 3.14 \times (3.173)^2 \times 0.280 = 8.852$

$S_t = 9{,}000/(2 \times 8.852\ ) = 508$ psi

finally S_E can be calculated:

$$S_E = \sqrt{(22{,}846)^2 + 4 \times (508)^2} = 22{,}868 \text{ psi}$$

Comparing this S_E to the S_A for most moderate temperature refinery services, this example would be operating at about 75% of the allowable stress range.

The effective section modulus was introduced to the B31 codes by Code Case 53 which follows.

Interpretations of Code For Pressure Piping

Case 53 (Reopened) - Stress-Intensification Factor

Inquiry*:* ASA B31.1-1955 (Par. 621d and Fig. 3.4), and ASA B31.3-1962 (Par. 319.3.6, and Table 319.3.6) provide a direct method for computing stress-intensification factors for full-size tees and fabricated branch connections. Application of the same factor to reducing-outlet connections, as recommended as a pro tem solution in Footnote 6, is believed to lead to gross over-evaluation of the stress range. Clarification and relief from an apparently unnecessarily severe requirement are desired.

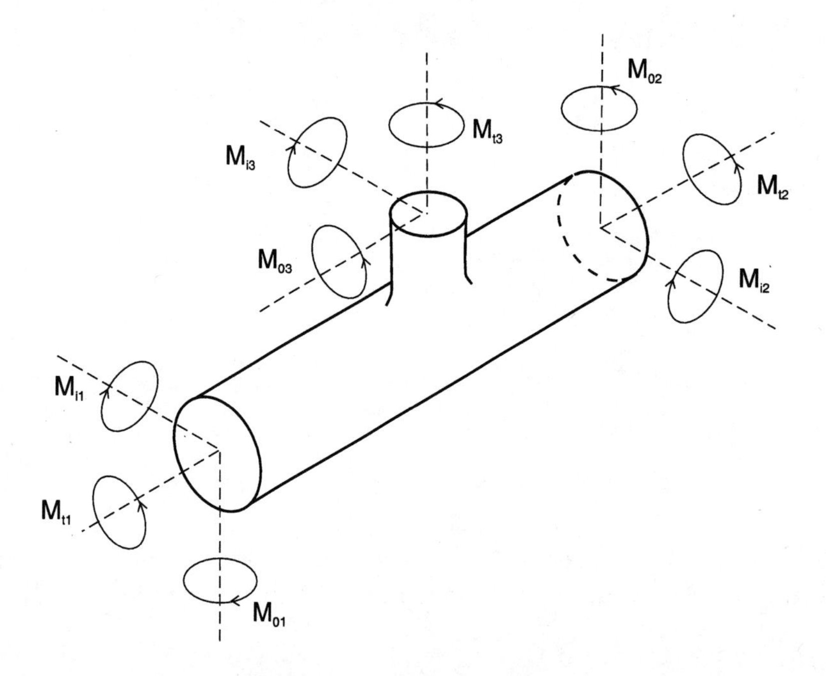

Figure 3.4 Applied moments.

Reply: Footnote 6 of Fig. 3.3 and Table 319.3.6 mentioned in the inquiry reflect lack of theoretical or experimental data at the time of its formulation. Isolated test results which have since become available warrant modifying the present rules for full-size tees and branch connections and extending them to cover reducing-outlet tees and branch connections as follows by reference to ASA B31.1-1955 and ASA B31.3-1962:

Determine in-plane bending moment M_i, out-of-plane bending moment M_o, and torsional moment M_t at the branch junction for each of the three legs and combine the resultant bending stress S_b and torsional stress S_t by Equation (13) in Par. 622(b), where:

$$S_t = \frac{M}{2Z} \text{ for header and branch} \qquad (53\text{-}1)$$

$$S_b = \frac{\left[(i_i M_i)^2 + (i_o M_o)^2\right]^{1/2}}{Z} \text{ for header (legs 1 and 2)} \qquad (53\text{-}2)$$

$$S_b = \frac{\left[(i_i M_i)^2 + (i_o M_o)^2\right]^{1/2}}{Z_e} \text{ for branch (legs 3)} \qquad (53\text{-}3)$$

The value i_o of the out-of-plane stress-intensification factor appearing in Equations (53-2) and (53-3) equals the value i presently computed using the dimensions of the pipe matching the run of a tee or the header pipe[1]. The value i_i of the in-plane stress-intensification factor is modified:

$$i_i = 0.75\, i_o + 0.25 \qquad (53\text{-}4)$$

The section modulus Z in Equations (53-1) and (53-2) is the section modulus of the header or branch pipe, for whichever the stress is being calculated. The effective branch section modulus in bending Z_e used in Equation (53-3) is a fictitious value used for purposes of test correlation:

$$Z_e = \pi r_2^2 T_s \qquad (53\text{-}5)$$

Where

r_2 = mean branch cross-sectional radius

T_s = lesser of $\bar{T}_h$ and $i_o \bar{T}_b$ = effective branch wall thickness

(The effective branch wall thickness was later changed to the lesser of $\bar{T}_h$ and $(i_i \bar{T}_b)$)

$\bar{T}_h$ = thickness of pipe matching run of tee or header exclusive of reinforcing elements

$\bar{T}_b$ = thickness of pipe matching branch

1. Note that pad or saddle thickness T_r should not be taken as greater than $1.5\,\bar{T}_h$ in formula for h for pad or saddle reinforced tee; this limitation was inadvertently omitted from the Code.

Stress Intensification Factor

In the preceding examples, the stress intensification factor (SIF) was calculated [B31.3 Appendix D] and used in the equation to determine the bending stresses resulting from thermal expansion or contraction. The SIF is an intensifier of the bending stresses local to a piping components such as tees or elbows and has a value equal to one (1.0) or greater. Each piping component is represented in this bending stress equation by its own SIF which is unique for the component. Components with low SIF's, (in the range of 1 or 2), because of their geometry containing smooth transition radii, have the greatest efficiency in blending bending stresses from one section of the piping through the component to the adjoining piping section. Components with sharp geometrical changes, such as an unreinforced fabricated tee, will have a high SIF in the range of 4 or 5, because the unreinforced tee will have a much lesser efficiency in blending bending stresses because of their sharp corner geometry.

SIF equations were first introduced into the piping codes in 1955. These equations were based on an extensive cyclic fatigue testing program conducted by A. R. C. Markl, H.H. George, and E.C. Rodabaugh at Tube Turns in the late 1940's and early 1950's. Figures 3.5 and 3.6 are examples of the fixtures used for SIF testing. This testing program first established an equation to represent the fatigue life of a butt weld in a straight length of pipe when cycled at a constant displacement. This equation, for ASTM A 106 Grade B piping material, is:

$$S = \frac{245{,}000}{N^{0.2}}$$

where "S" is the bending stress caused by the constant alternating displacement and "N" is the number of full displacement cycles until failure.

The SIF of the girth butt weld was assigned a value of one (1.0).

Figure 3.5 Stress intensification factor test equipment.

Figure 3.6 Testing engineers observe displacement indicator during SIF test.

This testing program then tested tees at the same displacement as the butt welded pipe samples and found that the tees failed after fewer displacement cycles. It was reasoned that the presence of the tee in the system intensified the stress to cause the earlier failure. The stress was calculated and compared to that of the butt welded pipe. The ratio of these stresses is the SIF of the tee, and the fatigue equation was modified for the tee and all other piping components as:

$$iS = \frac{245{,}000}{N^{0.2}}$$, where "i" is the SIF.

All piping components are represented in a flexibility analysis with a SIF of a value of 1.0 or greater as a representation of the component fatigue endurance as compared to that of the straight pipe with a butt weld.

Figure 3.7 compares the SIF of some common tee geometries. Figure 3.8 and Table 3.1 are tabulated expansion stresses for several tee geometries for the piping layout shown. The effects of the changing SIF caused by changing the tee geometry is very apparent in each of the *in-plane* and *out-plane* loading cases.

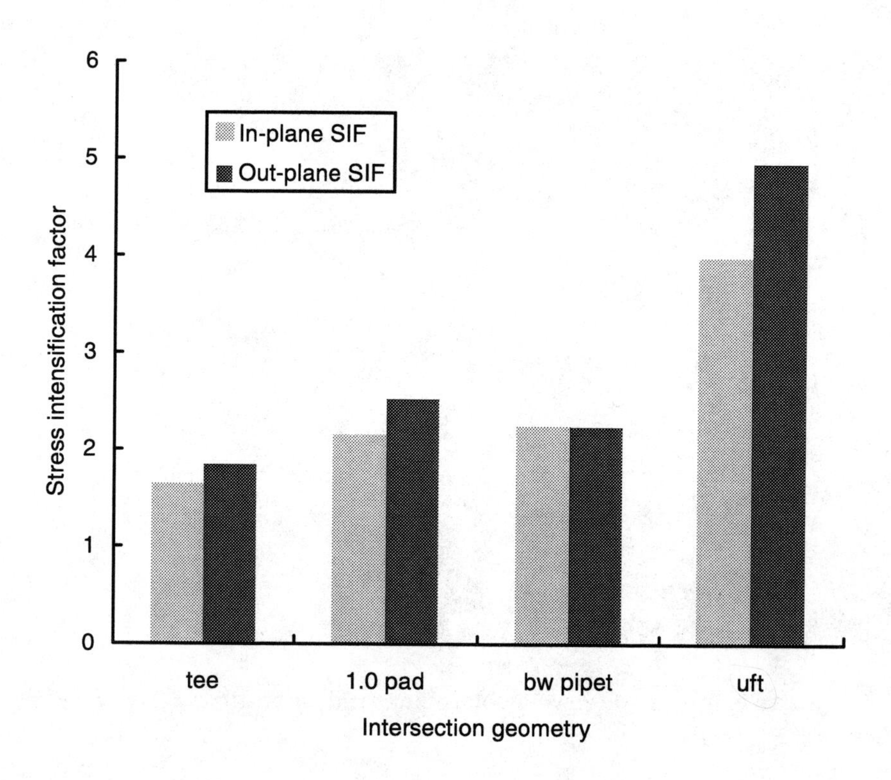

Figure 3.7 Comparison of common branch intersection stress intensification factor values.

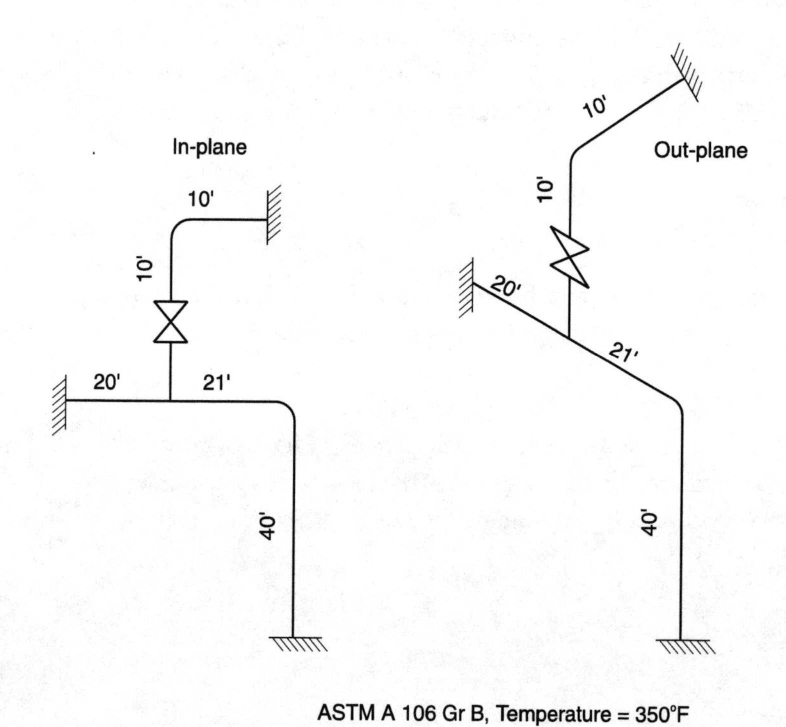

Figure 3.8 Three-anchor piping configurations.

The calculated expansion stress, S_E at an intersection in a piping system can change significantly with the selection of the type of branch intersection. Tabulated in Table 3.1 is the stress at the intersection for each of several intersection geometries for both in-plane and out-plane loadings.

Table 3.1 Calculated S_E for Various Tee Intersections

Item	Intersection Geometry	Expansion Stress, psi	
		In-Plane	Out-Plane
1	Welding Tee per ANSI B16.9	17,637	4,974
2	Weld-in contour insert	17,637	4,974
3	1.0 Pad reinforced fabricated Tee	23,269	6,686
4	Branch weld-on fitting	24,695	6,361
5	Extruded welding Tee, $r_x = 1.5$	37,911	11,159
6	Extruded welding Tee, $r_x = 1.0$	39,530	11,655
7	Extruded welding Tee, $r_x = 0.5$	41,361	12,215
8	Unreinforced fabricated Tee	43,402	12,840

Displacement Stresses of Dissimilar Welded Pipe Joint

When two different pipe materials having different thermal expansion coefficients are welded together for service in a hot (or cold) piping system, differential radial thermal expansion may occur. This differential thermal expansion will introduce a secondary stress at the point of weld. B31.3 does not provide a method for calculating this stress; however, piping designers must determine if this differential radial thermal expansion will cause over-strain (over stress) of the welded joint.

An example of a procedure for calculating the displacement stresses in a welded joint of dissimilar metals follows:

A thermal flexibility analysis at 1,000°F has revealed S_E = 8,000 psi at a butt-weld joining a process pipe to a furnace tube (Figure 3.9). The process piping material is ASTM A 335 Grade P22; the furnace tube material is ASTM A 312 Grade TP304H. Is the butt-weld joint overstressed? (The installed temperature is 70°F, ΔT = 1000 - 70 = 930°F.)

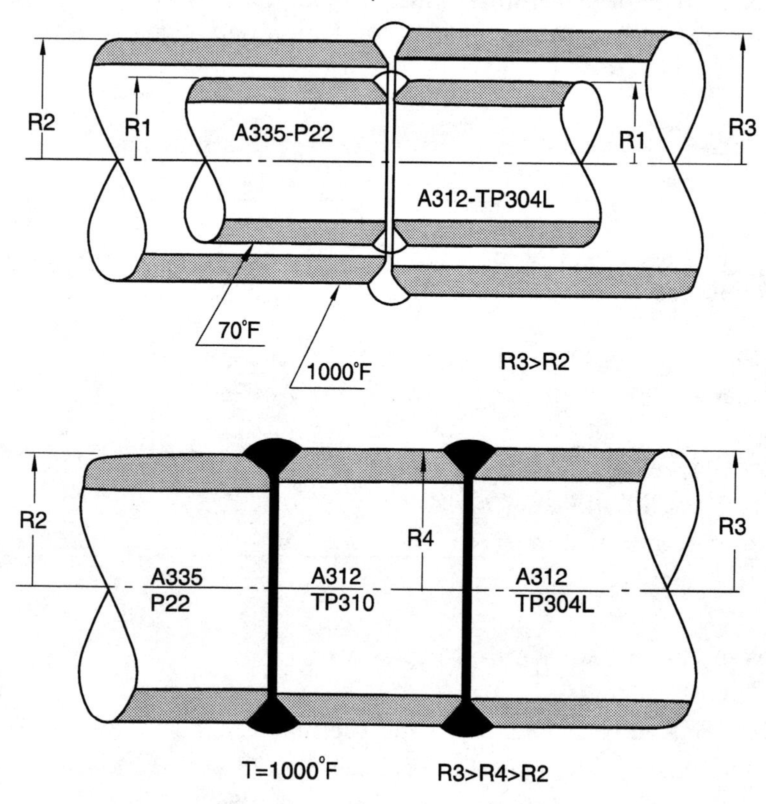

Figure 3.9 Radial thermal expansion at dissimilar weld joint.

Solution: Using Equation 1a (see Allowable Stress Range section in this chapter), the S_A at 1,000°F for each material, the cold modulus of elasticity Ec, and the expansion coefficient $\Delta\alpha$ are shown in Table 3.2.

Table 3.2 Values For σ Calculations

	Pipe A 335-P22 (2¼ Cr-1Mo)	**Tube** A 312-TP304H (18Cr-8Ni)	**Code Reference**
S_c	20,000 psi	20,000 psi	(B31.3 Appendix A-1)
S_h	7,800 psi	13,800 psi	(B31.3 Appendix A-1)
S_A	26,950 psi	28,400 psi	(B31.3 Eq. 1a)
Ec	30.6×10^6 psi	28.3×10^6 psi	(B31.3 Table C-6)
$\Delta\alpha$	7.97×10^{-6} μin./in.	10.29×10^{-6} μin./in.	(B31.3 Table C-3)

The radial thermal expansion stress at dissimilar metal joints is calculated as follows;

$$\sigma = 0.5Ec\Delta T\Delta\alpha < S_A - S_E$$

Assuming Ec is the same for both materials, and the pipe and furnace tube have the same wall thickness at the weld location, then the equation can be rearranged to solve for strain and allowable strain;

$$\Delta T\Delta\alpha < 2\,(S_A - S_E)/Ec$$

Solving our problem:

Pipe strain, $\Delta\alpha\Delta T = 7.97 \times 10^{-6} \times 930 = 7.4 \times 10^{-3}$ in./in.
Tube strain, $\Delta\alpha\Delta T = 10.29 \times 10^{-6} \times 930 = 9.6 \times 10^{-3}$ in./in.

Then the differential strain at the point of weld is $\Delta T\Delta\alpha = (9.6 - 7.4)10^{-3} = 2.2 \times 10^{-3}$ in./in. strain and the allowable strain is $2(S_A - S_E)/Ec = 2\,(26{,}950 - 8{,}000)/30.6 \times 10^{6} = 1.238 \times 10^{-3}$ in./in. strain.

The calculated strain of 2.2×10^{-3} exceeds the allowable strain of 1.238×10^{-3}. The joint is overstressed. A transition piece made of a material with an intermediate expansion coefficient must be added between the pipe and tube to reduce this stress (strain).

Try ASTM A 312 Grade TP310, a 25Cr-20Ni material, $\Delta\alpha = 9.18 \times 10^{-6}$in./in.°F, (B31.3 Table C-3), and $\Delta\alpha\Delta T = 8.5 \times 10^{-3}$ in./in.

The strain between transition piece and the process pipe is:

$$(8.5 - 7.4)10^{-3} = 1.1 \times 10^{-3}$$

The strain between the furnace tube and the transition piece (ASTM A 312 Grade TP310) is:

$$(9.6 - 8.5)10^{-3} = 1.1 \times 10^{-3}$$

Both strains are within the allowable strain limit. Therefore, a transition piece of 25Cr-20Ni, approximately 6 to 8 inches long is to be welded between the ASTM A 312 Grade TP304H and the ASTM A 335 Grade P22 materials.

The $\Delta\alpha\Delta T$ can be graphically represented as shown in Figure 3.10 for the materials. Differences in expansion strain at temperature can be extracted.

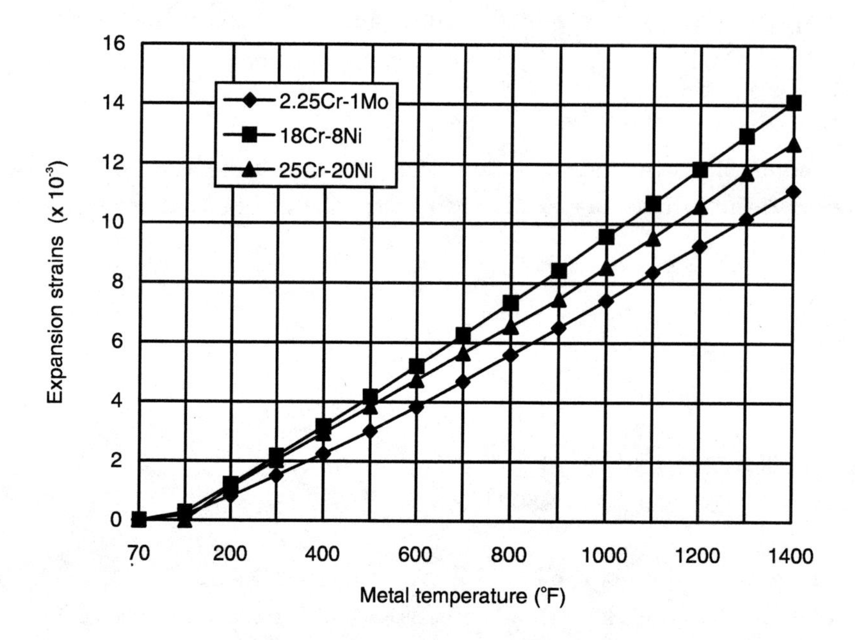

Figure 3.10 Values of $\Delta\alpha\Delta T$.

Cold Spring

Cold spring [¶319.5.1] in a piping system is the intentional deformation of the piping for the purpose of reducing pipe end reactions on supporting steel or equipment. This deformation is introduced during fabrication and erection by cutting the pipe length long or short, depending on the expected thermal expansion. Piping systems operating above the installed temperature would be cold sprung by shortening the pipe length by an amount equal to or less that the expected thermal expansion. To illustrate this concept, consider the following:

A designer has located an expansion loop in a 500°F carbon steel steam line in the pipe rack. Assume the total thermal expansion from the installed temperature of 70°F is calculated to be 4 inches (length of pipe between anchors times the material expansion coefficient found in B31.3 Table C-1). Further assume that the thermal force on the anchors caused by the expansion and loop leg deflection is 5,000 pounds. An analysis of the anchor steel indicated the steel is overloaded. The designer then elects to install 50% cold spring where the pipe length between the anchors will be cut short by 50% of the total thermal displacement or 2 inches. The thermal reaction on the anchor steel is then reduced to a lower value, R_m, by the equation:

$$R_m = R\left(1 - \frac{2C}{3}\right)\frac{E_m}{E_a}$$

where

R = reaction force from thermal analysis (5,000 lb for our example)
E_m = modulus of elasticity at the maximum temperature (see B31.3 Table C-6)
E_a = modulus of elasticity at the installed temperature (see B31.3 Table C-6)
C = cold spring factor varying from 0 for no cold spring to 1.0 for 100% cold spring.

The constant ⅔ is an uncertainty factor, since difficulty may arise in actually obtaining the desired cold spring.

Then for our illustration,

$$R_m = 5{,}000\left(1 - \frac{2(0.50)}{3}\right)\frac{27.3}{29.5} = 3085 \text{ lb.}$$

The design anchor load on the steel is reduced by 38% (from 5,000 to 3,085 lb) when 50% cold spring is implemented.

Cold spring can extend to a value of 100%. If this was the case in the above example, R_m would be reduced even more.

For cryogenic or cold pipe service, where E_m is greater than E_a, the thermal reactions will increase when this R_m equation is used and the thermal analysis is conducted using E_a.

As seen in this example and using the R_m equation, cold spring can be used to reduce reactions. Cold spring can not be used to reduce thermal stress range. S_E can not be lowered by the presence of cold spring. The range of thermal displacement cycling will remain unchanged regardless of cold spring as graphically illustrated in Figure 3.11. The stress range remains the same regardless of cold spring.

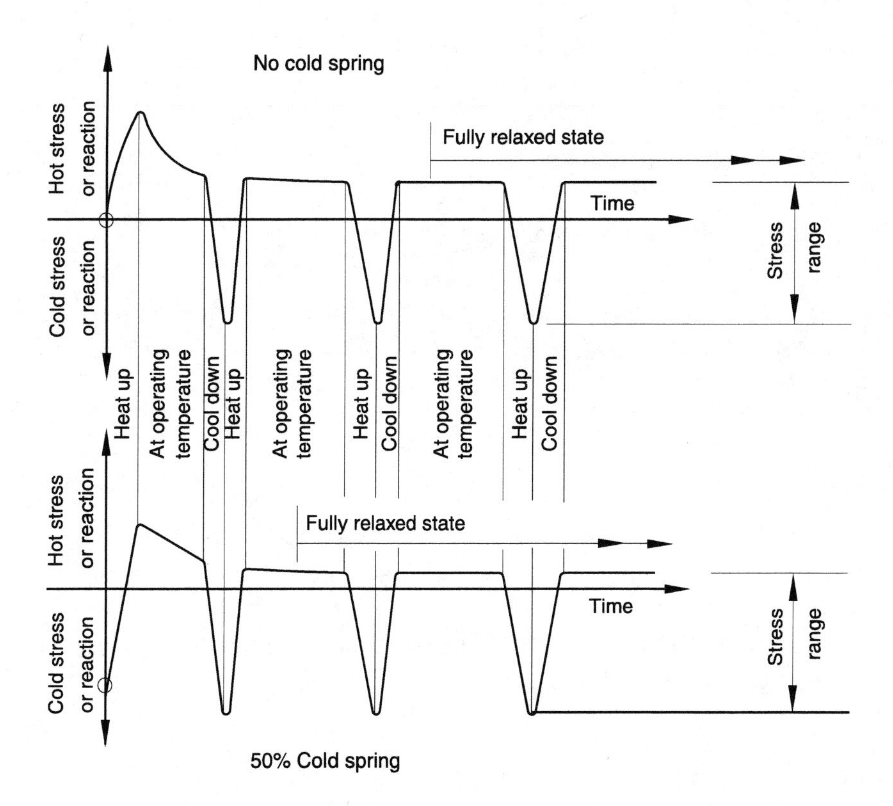

Figure 3.11 Effects of relaxation upon reactions and stresses, stress range is unchanged.

Sustained Load Stress

Sustained load stresses are primary stresses caused by pressure or weight and will not diminish with time or as local yielding of the stressed pipe occurs.

B31.3 establishes limits for sustained load stresses. In the Design Conditions section found in Chapter 2, the first sustained stress limit is listed: "The nominal pressure stress shall not exceed the yield strength of the material at temperature" [¶302.2.4(b)].

The nominal pressure stress is the hoop stress, sometimes called the circumferential pressure stress. The hoop stress, σ_h, is calculated as follows using the *thin wall formula*:

$$\sigma_h = \frac{PD}{2t} \text{ psi}$$

where

P = internal pressure, psig
D = pipe outside diameter, in.
t = pipe nominal wall thickness less corrosion, erosion, and mechanical allowances, in.

Example: Find the hoop stress in a 12.75 in., 0.375 in. wall thick pipe, ASTM A 106 Grade B seamless, at 500 psig internal pressure. The pipe service requires a corrosion allowance of 0.0625 in.

Solution: The nominal wall of the pipe is 0.375 in. The mill under-run tolerance for seamless pipe is 12.5%; therefore, the wall thickness to be used is:

$$t = 0.375\ (1 - 0.125) - 0.0625 = 0.2656$$

then $\sigma_h = 500 \times 12.75/(2 \times 0.2656) = 12{,}000$ psi

This pipe wall thickness is adequate to protect against a primary stress failure of bursting, provided the yield strength of the pipe material is greater than 12,000 psi at the temperature of the pressurized condition. (Note: The stress limit for pressure design is S_h, not the yield strength of the material.)

The second sustained load stress limit B31.3 presents is on longitudinal stresses due to sustained loads, S_L [¶302.3.5(c)]. These are stresses that are directed along the axis of the stressed pipe (tensile or compressive) and are caused by pressure, weight, and other sustained loads.

B31.3 does not provide an equation for calculating S_L. Consideration is being given by B31.3 to adopt the B31.1 code S_L equation. The B31.1 equation is :

$$S_L = \frac{PD}{4t} + \frac{0.75iM_A}{Z} \leq 1.0\ S_h$$

NO "MT"

where

- t = nominal wall less corrosion, erosion allowance, inches, mill tolerance is not removed.
- M_A = resultant moment due to weight and other sustained loads, in.-lb.
- i = stress intensification factor for the component under analysis. Note, B31.1 has only one SIF where B31.3 has two, in-plane and out-plane. If the designer wishes to adopt the B31.1 S_L equation for B31.3 analysis, then the SIF to be selected would be the greater of i_i or i_o.

An example of the B31.3 implied S_L approach is as follows.

The longitudinal stress due to pressure, S_{LP} is calculated as:

$$S_{LP} = \frac{PD}{4t}$$

MILL TOL. CAN BE IGNORED

The terms of this equation are the same as for σ_h except mill under-run tolerances are not to be removed from t. The value of t used in the calculation of S_{LP} is the nominal wall thickness less mechanical, corrosion, and erosion allowances, $\overline{T}$ - c.

The weight stress contribution to this S_L is longitudinal bending stresses caused by pipe sag, pipe overhang, or any other bending caused by gravity. The bending stress caused by weight is calculated as:

$$S_{WL} = \frac{WL}{Z}$$

where

- W = weight of the overhang, concentrated at the center of gravity of the system, lb
- L = length from point of support to W, in.
- Z = section modulus of pipe, in.^{3}

Example: Calculate the S_L of a 12.75 in. outside diameter, 0.375 in. nominal wall, seamless pipe with the following conditions.

P = 650 psig; ca = 0.0625 in.; and the center of gravity load of 400 lb is located 10 ft from the pipe support.

Solution:

$t = 0.375 - 0.0625 = 0.3125$ in.

O.D. = 12.75 in.

I.D. = $12.75 - 2 \times 0.3125 = 12.125$ in.

$Z = \pi(12.75^4 - 12.125^4)/(32 \times 12.75) = 37.06$ in.^{3}

then $S_{LP} = 650 \times 12.75/(4 \times 0.3125) = 6{,}630$ psi

and $S_{LW} = 10 \times 12 \times 400/37.06 = 1{,}295$ psi

finally $S_L = S_{LP} + S_{LW} = 6{,}630 + 1{,}295 = 7{,}925$ psi

The allowable stress for S_L is S_h, the hot allowable stress of the material at temperature. The allowable stress for S_L in cold pipe service is S_c. The allowable stress for S_L in cold pipe service is the lesser of S_c or S_h.

Occasional Load Stresses

Occasional load stresses in piping systems are the sum of those stresses caused by loads such as relief valve discharge, wind or earthquake [¶302.3.6]. These stresses are calculated considering:

a) the pipe deflection caused by wind load, acting as a horizontal constant pressure on the outside surface of the pipe, or
b) the pipe deflection caused by earthquake loads, acting as a horizontal or vertical acceleration of the mass or weight of the piping system. The typical method of analysis for earthquake loadings, in a location subject to a horizontal acceleration of 0.28 G, for example, is to determine the stresses resulting from deflections caused by a horizontal constant force equal to 28% of the pipe weight applied in the same horizontal (or vertical) manner as a wind load.

The allowable stress for occasional loads, S_{OL}, summed with the stresses due to sustained loads, S_L, is $1.33S_h$.

$$S_{OL} + S_L \leq 1.33S_h$$

Wind and earthquake need not be considered as acting concurrently.

Wind Loads

B31.3 directs the designer to use the method of analysis stated in ASCE 7-93 [¶301.5.2] for the determination of wind loads based on exposure categories. Wind loads, W_{WL}, are calculated by the equation (**Note**: The current Code edition references ASCE 7-88. This will soon change to ASCE 7-93.):

$$W_{WL} = q_z G_z C_f A \text{ lb}$$

The equation for the calculating q_z, the velocity pressure, at height z is presented as:

$$q_z = 0.00256K_z(IV)^2 \text{ lb/ft}^2$$

where

K_z= velocity pressure exposure coefficient (partial data duplicated in Table 3.4 following, values listed for Exposure "D" only)
I = importance factor (partial data duplicated in Table 3.5)
V = wind speed, miles/hour from ASCE 7-93 (see Figure 3.12)
G_z= gust response factor (partial data duplicated in Table 3.4)
A = area of pipe surface, including insulation exposed to wind, ft^2

ASCE 7-93 lists four exposure categories, describing the types of terrain for wind load calculations. Examples of these terrains are:

Exposure A - large city centers.
Exposure B - urban and suburban areas, wooded areas.
Exposure C - open terrain with scattered obstructions having heights less than 30 feet.
Exposure D - flat, unobstructed areas exposed to wind flowing over large bodies of water.

Table 3.3 Building and Structure Classification

Nature of Occupancy	**Category**
All buildings and structures except those listed below.	I
Buildings and structures where the primary occupancy is one in which more than 300 people congregate in one area.	II
Essential facilities such as hospitals, fire or rescue and police stations, power stations, and other utilities required in an emergency.	III
Buildings and structures that represent a low hazard to human life in the event of a failure.	IV

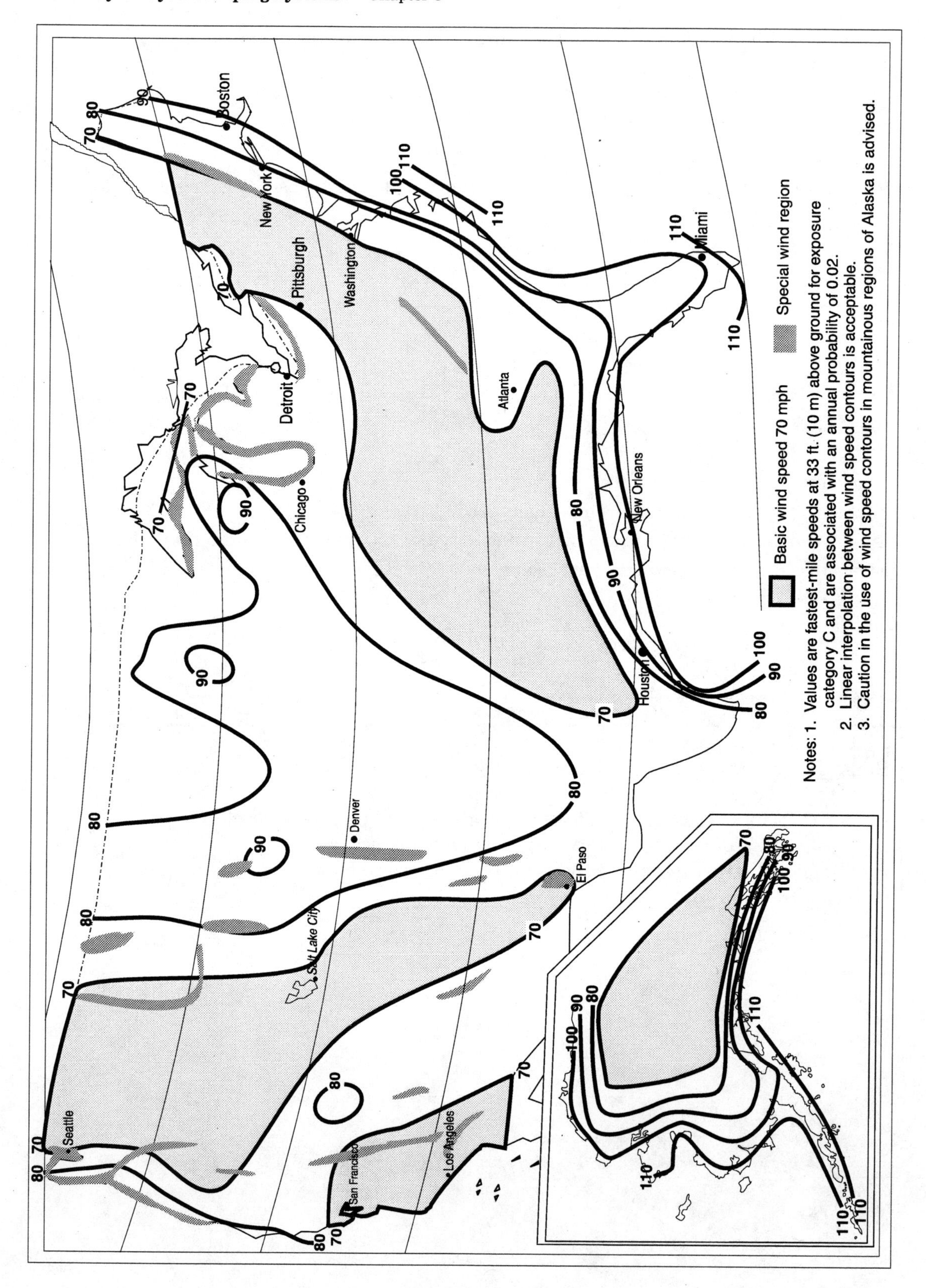

Figure 3.12 Basic wind speeds, miles per hour.

Table 3.4 Velocity Exposure Coefficient, K_Z, and Gust Response Factor, G_Z, for Exposure D

Height Above Ground Level, z (ft)	K_Z	G_Z
0 - 15	1.20	1.15
20	1.27	1.14
25	1.32	1.13
30	1.37	1.12
40	1.46	1.11
50	1.52	1.10
60	1.58	1.09
70	1.63	1.08
80	1.67	1.08
90	1.71	1.07
100	1.75	1.07
120	1.81	1.06
140	1.87	1.05
160	1.92	1.05
180	1.97	1.04

The importance factors listed are for essential facilities, depending upon the category listed in Table 3.3.

Table 3.5 Importance Factor, I

	Importance Factor, I	
Category	**100 miles from hurricane ocean line**	**at hurricane ocean line**
I	1.00	1.05
II	1.07	1.11
III	1.07	1.11
IV	0.95	1.00

An example of the calculation of the *velocity pressure*, q_z, follows:

Find q_z for a pipe length 140 feet above the ground in a 100 mile/hour wind speed at a Category I, Exposure D structure classification plant on the Texas Gulf Coast.

Solution:

K_z = 1.87, from Table 3.4
I = 1.05, from Table 3.5
V = 100 mi/hr
$q_z = 0.00256 \times 1.87 \times (1.05 \times 100)^2$
q_z = 52.78 lb/ft^2

The area of the pipe over which q_z acts is determined by:

Area = pipe length x pipe width x C_f

C_f is a force coefficient and is based on the shape and height of the structure being acted on by the wind. For pipe, C_f is determined from Table 3.6 (h is the height of the pipe in feet above the ground and D is the pipe diameter, including insulation thickness, in feet, q_z, velocity pressure, lb/ft^2).

Table 3.6 C_f, Force Coefficients (for Pipe)

	C_f for h/D Values		
Pipe with	**1**	**7**	**25**
$D\sqrt{q_z} > 2.5$	0.5	0.6	0.7
$D\sqrt{q_z} \leq 2.5$	0.7	0.8	1.2

Returning to the q_z example, now find the wind load, W_{WL} on a 25 ft horizontal length of 8.625 in. outside diameter, schedule 40 pipe with 4 in. of insulation.

q_z = 52.78 lb/ft^2

C_f = 0.7, from Table 3.3

A = 25 (8.625 + 2 x 4)/12 = 34.63 ft^2/25 ft of length

The gust response factor, G_z = 1.05 from Table 3.4.

Finally, the wind load on the 25 foot length of pipe and insulation is:

W_{WL} = 52.78 x 1.05 x 0.7 x 34.63 = 1,343 lb

Figure 3.13 illustrates the change in wind pressure expressed in lb/ft^2 of pipe as the wind velocity and pipe height above the ground change.

The occasional load stress, σ_{OL}, caused by this wind load can be calculated using beam equations. The following is an example of the calculation of σ_{OL} using the equation for a simply supported (and guided) beam with the bending moment caused by the uniform wind load, $M = W_{UWL} \times L^2/8$ inch-pound. The occasional load stress for the twenty five foot long section of NPS 8 inch pipe in the previous wind load example can be determined as follows:

The uniform load, W_{UWL}, is from the wind load of:

$W_{UWL} = 52.78 \times 1.05 \times 0.7 \times (8.625 + 8)/12 = 53.74$ lb/ft
$W_{UWL} = 4.5$ lb/in.

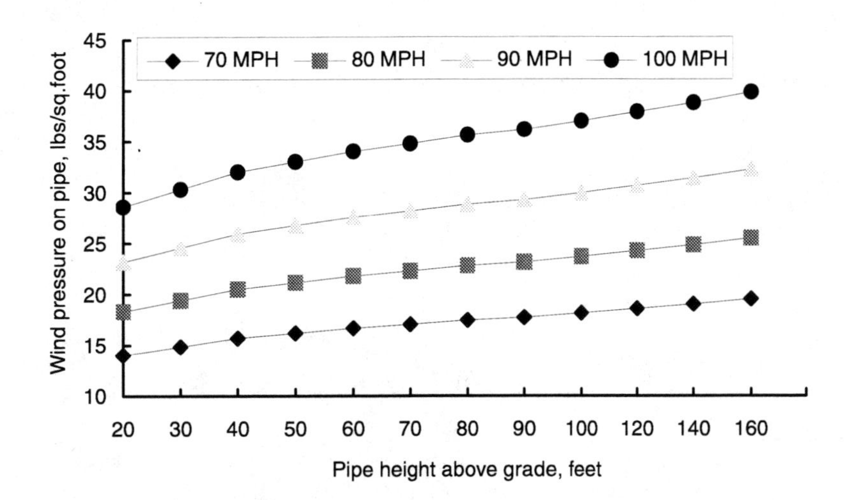

Figure 3.13 Wind pressure on pipe.

The mid-span stress, σ_{OL} is calculated by, $\sigma_{OL} = M/Z$ psi, where

$M = W_{UWL}L^2/8 = 4.5 \times (25 \times 12)^2/8 = 50{,}625$ in.-lb
$Z = 16.8$ in.^{3}
$\sigma_{OL} = 50{,}625/16.8$
$\sigma_{OL} = 3{,}013$ psi

This stress is then added to the longitudinal stresses due to sustained loads, S_L, and this summation must be less than the allowable stress of $1.33S_h$ [¶302.3.6], ($1.33S_c$ for cold pipe service).

Note that $1.33S_h$, where S_h = ⅔ yield strength of the material, is equal to about 0.9 yield. These sustained stresses are primary stresses that shall not exceed the yield strength of the material at temperature. This is another example of the maximum principal stress failure theory.

Earthquake

B31.3 directs the piping designer to ASCE 7-93 [¶301.5.3] for the procedure to determine the forces acting on pipe caused by earthquake ground motions. These forces which act in both the horizontal and vertical directions will cause displacements in the piping similar to wind. The designer is to determine the resulting occasional load stress from these displacements, add the sustained load stresses, S_L for comparison to the allowable stress of $1.33S_h$ [¶302.3.6].

The ASCE 7-93 equation to determine earthquake forces on pipe is:

$$F_p = A_v C_c P a_c W_c \text{ lb}$$

where

A_v = the seismic coefficient representing effective peak velocity-related acceleration from Figure 3.14
C_c = the seismic coefficient for pipe from Table 3.7
P = the performance criteria factor from Table 3.7
a_c = the amplification factor from Table 3.9
W_c = operating weight of the pipe, lb

Table 3.7 Seismic Coefficients C_c and Performance Criteria Factor P

	Seismic Coefficient	**Performance Criteria Factor "P" Seismic Hazard Exposure Group**		
Pipe system:	"Cc"	I	II	III
Gas and high hazard	2.0	1.5	1.5	1.5
Fire suppression	2.0	1.5	1.5	1.5
Other pipe systems	0.67	N/R*	1.0	1.5

*N/R = not required

See Table 3.8 for seismic hazard exposure group.

Table 3.8 Seismic Hazard Exposure Group

Structure Description	**Group**
Structures not assigned to Group II or III	I
Buildings that have a substantial public hazard due to occupancy or use including schools, medical facilities, meeting facilities, power generating stations, and other public facilities not in Seismic Group III	II
Fire or rescue and police stations, hospitals, power generating stations required as emergency back-up	III

Table 3.9 Attachment Amplification Factor (a_c)

Component Supporting Mechanism	Attachment Amplification Factor (a_c)
Fixed or direct connection	1.0
Seismic-activated restraining device	1.0
Resilient support system where	
$T_c/T < 0.6$ or $T_c/T > 1.4$	1.0
$T_c/T \geq 0.6$ or $T_c/T \leq 1.4$	2.0

T_c is the fundamental period of the pipe and its pipe support attachment to the supporting structure and T is the fundamental period of the supporting structure.

An example of the application of the ASCE 7-93 seismic procedure follows:

Find the horizontal seismic design force, F_{ph} for a 30 foot length of hazardous gas pipe weighing 1,000 pounds that is rigidly guided and supported at each end in a plant located on the west coast of California.

Solution:

$A_v = 0.4$ Figure 3.14
$C_c = 2.0$ Table 3.7
$P = 1.5$ Table 3.7
$a_c = 1.0$ Table 3.9
$W_c = 1{,}000$ lb

$F_{ph} = 0.4 \times 2.0 \times 1.5 \times 1.0 \times 1{,}000 = 1{,}200$ lb

The horizontal seismic force of 1,200 pounds calculated for this 30 foot length of pipe must be restrained by the pipe guides. The occasional load stress resulting from this horizontal force can be calculated by the same uniformly loaded, simply supported beam equations as used in the wind load example. To illustrate this stress calculation, assume the gas pipe to be NPS 8 schedule 60 pipe, (30 ft long).

The mid span stress is calculated by $\sigma = M/Z$ psi.

The uniform horizontal load, UHL = 1,200/(30 x 12) = 3.33 lb/in.

The moment, M, at mid span = $[UHL \times L^2] \div 8 = [3.33 \times (30 \times 12)^2] \div 8 = 53{,}946$ in.-lb

$Z = 20.6$ in.^{3}

Then $\sigma = 53{,}946/20.6 = 2{,}618$ psi

This occasional load stress is added to the sustained load stress, S_L and compared to the allowable stress, $1.33S_h$ [¶302.3.6].

The ASCE 7-93 earthquake procedures above can be used to calculate the vertical seismic force, F_{pv}, by using ⅓ of the horizontal C_c factor in the F_{ph} equation above. All other term values remain the same as for the horizontal force calculation. The vertical seismic force for the above example is calculated as follows:

$$F_{pv} = 0.4 \text{ x } (1/3) \text{ x } 2.0 \text{ x } 1.5 \text{ x } 1.0 \text{ x } 1{,}000 = 400 \text{ lb}$$

The stress resulting from F_{pv} is negligible as compared to that resulting from F_{ph} and would not be added to the horizontal stress value but is considered separately (as not acting at the same time). This F_{pv} load is considered in the selection of pipe supports; hold-down supports may be required if F_{pv} exceeds the weight of the pipe.

Figure 3.14 Seismic effective peak velocity-related acceleration (A_V).

Safety Relief Valve Discharge [¶301.5.5]

The occasional load stresses resulting from a safety relief valve discharge occur at the intersection of the relief valve branch pipe and the header. It is at this location that bending moments caused by the discharge thrust will be the greatest and the resulting stress must be calculated. This stress calculation would be the same as the calculation of bending stress discussed earlier except that in the place of bending forces caused by thermal expansion, the forces of the relief valve thrust are used to generate the bending moments. The allowable stress for this bending stress is $1.33S_h$. This force can be calculated using the force calculation procedure for open- and closed-discharge systems in API-RP520.

An example of this relief valve reaction and resulting bending stress calculation follows:

A low pressure steam pipe relief valve discharges to the atmosphere, open discharge, due to an over-pressure condition. What is the reaction force, F, and the bending stress at the relief valve pipe intersection to the run pipe? Assume there is no support on the relief valve tail pipe. The designer has selected a safety relief valve vesselet,[1] a welded-in contoured insert branch fitting to join the relief valve inlet pipe to the run (see Figure 3.15). The run pipe is 12.75 inches outside diameter, with 0.375 inch nominal wall thickness. The branch pipe to the relief valve is 4.50 inches outside diameter, with 0.237 inch nominal wall thickness. The terms below are defined in API RP-520 (see Figure 3.16).

Discharge conditions:

$W = 19{,}550$ lb/hr; $k = 1.32$; $T = 415°F = 875°R$; $M = 18$; $A_o = 28.9$ in.^{2}
$P_2 = 0$ (flow is not choked.)

$$F = \frac{W\sqrt{\dfrac{kT}{(k+1)M}}}{366}$$

(From API-RP520, F in pounds of thrust.)

$$F = \frac{19550\sqrt{\dfrac{1.32 \times 875}{2.32 \times 18}}}{366}$$

$F = 281$ lbf

[1] Manufactured by WFI International, Houston, Texas.

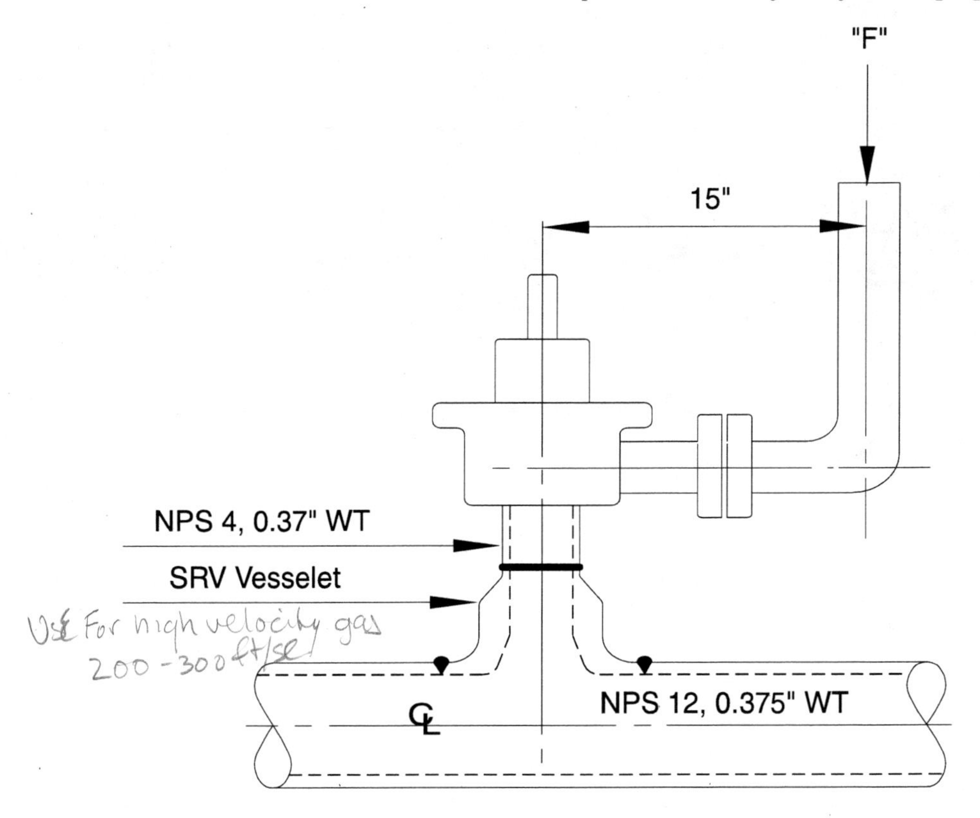

Figure 3.15 Safety relief valve installation.

Neither API nor B31.3, specifically recommends including a dynamic load factor for relief valve thrust calculations. However B31.1, the power piping code, recommends multiplying the force F by a 1.2 dynamic load factor to better approximate the instantaneous force at the time the relief valve opens. By employing this factor, the design force for the bending stress calculation is:

F = 1.2 x 281 = 337 lb

Assume the distance from the center line of the relief valve inlet pipe to the center line of the relief valve discharge pipe is 15 inches. (This is the moment arm to be multiplied by the force to produce the bending moment.) Then, the bending moment is:

M_i = 337 x 15 = 5,055 in.-lb moment.

Assume the relief valve is oriented for in-plane reaction piping layout as shown in Figure 3.15. The stress intensification factor for the insert fitting is (see B31.3 Appendix D):

$i_o = 0.9/h^{2/3}$; $h=4.4\bar{T}/r_2$; $r_2 = 0.5\ (12.75 - 0.375) = 6.1875$; $\bar{T} = 0.375$; $h = 0.266$
$i_o = 2.17$, and $i_i = 0.75 i_o + 0.25 = 1.88$

The effective section modulus, Z_e for the reducing intersection:

$Z_e = \pi r_2^2 T_s$, $r_2 = 0.5\ (4.50 - 0.237) = 2.1315$. Note: r_2 is the nominal radius of the branch pipe.
$T_s = 0.375$, then $Z_e = 5.352$

$$\text{Finally } S_b = \frac{\sqrt{(1.88 \text{ x } 5{,}055)^2}}{5.352} = 1{,}775 \text{ psi and}$$

$S_b = 1{,}775$ psi.

This S_b is compared to $1.33S_h$ to determine if the intersection is over-stressed.

API RP-520

2.4.1 Determining Reaction Forces In An Open-Discharge System
The following formula is based on a condition of critical steady-state flow of a compressible fluid that discharges to the atmosphere through an elbow and a vertical discharge pipe (see Figure 3.16). The reaction force (F) includes the effects of both momentum and static pressure; thus, for any gas or vapor,

$$F = W\frac{\sqrt{\frac{kT}{(k+1)M}}}{366} + (A_o \times P_2)$$

where

F = reaction force at the point of discharge to the atmosphere, in pounds (Newtons).
W = flow of any gas or vapor, lb/hr (kg/s).
k = ratio of specific heats (C_p/C_v).
C_p = specific heat at constant pressure.
C_v = specific heat at constant volume.
T = temperature at inlet, °R (°F + 460).
M = molecular weight of the process fluid.
A_o = area of the outlet at the point of discharge, in.^{2} (mm^2).
P_2 = static pressure at the point of discharge, in psi gauge (bar gauge).

2.4.2 Determining Reaction Forces In A Closed-Discharge System
Pressure relief valves that relieve under steady-state flow conditions into a closed system usually do not create large forces and bending moments on the exhaust system. Only at points of sudden expansion will there be any significant reaction forces to be calculated. Closed-discharge system, however, do not lend themselves to simplified analytic techniques. A complex time-history analysis of the piping system may be required to obtain in the true values of the reaction forces and associated moments.

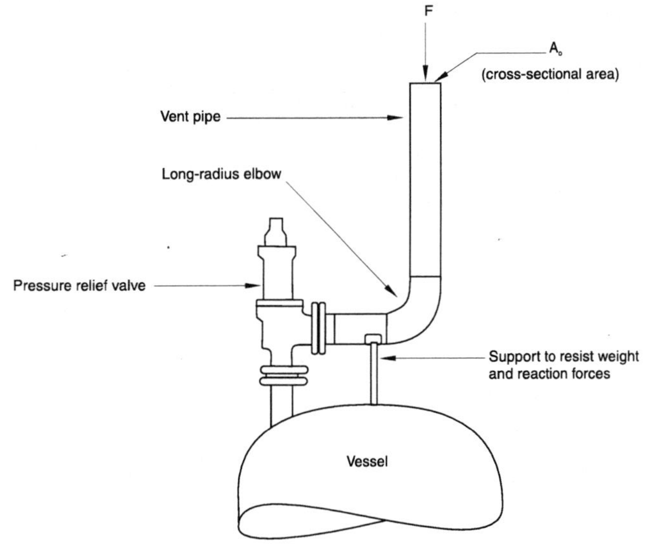

Note: the support should be located as close as possible to the center line of the vent pipe.

Figure 3.16 Open discharge piping with support.

Increasing Flexibility

B31.3 offers several methods to increase the flexibility [¶319.7] of a piping system. Added flexibility may be necessary to reduce the piping system loads on load sensitive equipment such as pumps, turbines, or compressors. The traditional method for increasing flexibility is to add expansion loops or off-sets in the piping layout. The key objective of adding flexibility by using loops or off-sets is to move the center of gravity of the system away from the line of thrust.

Consider a simple two anchor piping layout and draw a line of thrust connecting the two anchors (Figure 3.17). Estimate the center of gravity. Flexibility is increased when the added pipe moves the center of gravity away from this line of thrust.

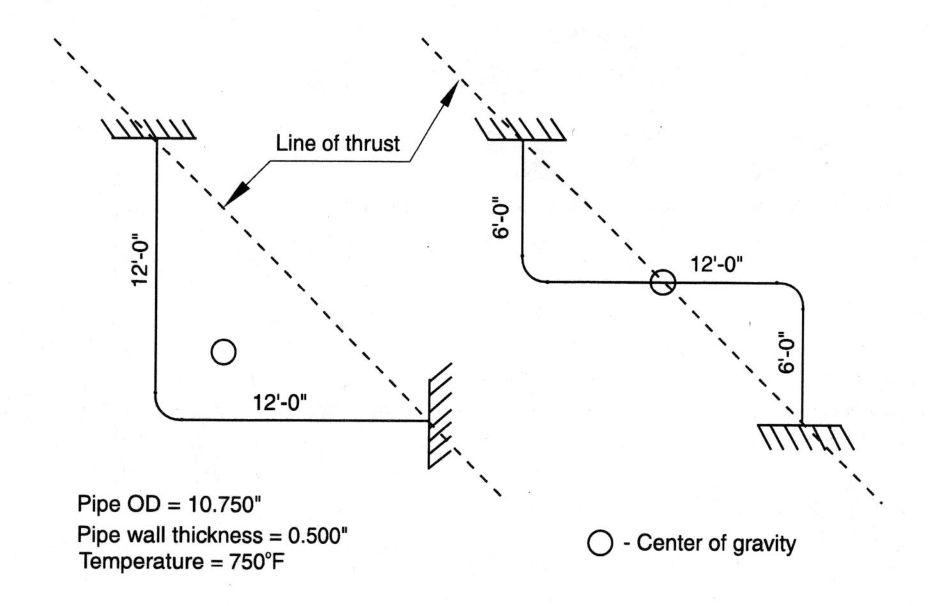

Figure 3.17 Increasing pipe flexibility.

This center-of-gravity/line-of-thrust concept is further illustrated by the following two computer analyses of the above pipe layouts (Figures 3.18 and 3.19). Both piping layouts are the same pipe size and temperature, and the anchors are the same distance apart. The "L" shape layout has a maximum expansion stress of 24,455 psi. The "Z" shaped has 42,594 psi. The L shape moved the center of gravity, "cg," away from the line of thrust which produced a lower stress and greater flexibility even though the Z shape had one more elbow.

CAESAR II (Ver. 2A) All Properties Listing (Pipe)
Job Description: B31.3 Sample #1

	From / To	X / Delta Y / Z	Dia. / Wall Thk. / Ins. Thk.	Temp 1 / Temp 2 / Temp 3	Pressure / 1 / 2	Elastic Mod. / Poissons R. / Corosion	Pipe D / Insul. D / Fluid D			
Bend	5.	.000	10.750	750.00000	.0	.2790E+08	.2899	Bend Radius = 15.000	Fitting Thk. = .5000	
	10.	-12.000 ft.	.500	.00000	.0	.292000	.0000			
		.000	.000	.00000	Mat #1	.000000	.0000	RSTR Node = 5.	DIR = A	CN = 0.
								STIF=.100000E+13	GAP = .0000	MU = .00000
								Code B31.3	SC = 20000.00	
								SH1 = 13000.00	SH2 = .00	SH3 = .00
								F1 = 1.00	F2 = 1.00	F3 = 1.00
Strt	10.	12.000 ft	10.750	750.00000	.0	.2790E+08	.2899	RSTR Node = 15.	DIR = A	CN = 0.
	15.	.000	.500	.00000	.0	.292000	.0000	STIF =.100000E+13	GAP =.0000	MU = .00000
		.000	.000	.00000		.000000	.0000			

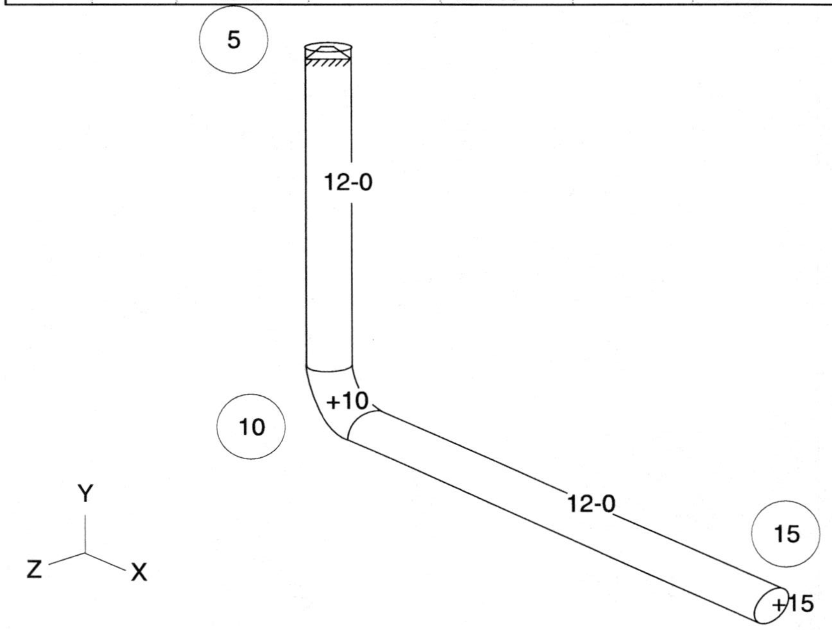

CAESAR II Displacement Report Problem #1
Case 3 (EXP) D3 (EXP) = D1-D2

	Translations (in.)			Rotations (deg.)		
Node	DX	DY	DZ	RX	RY	RZ
5	.0000	.0000	.0000	.0000	.0000	.0000
10	-.658	-.7115	.0000	.0000	.0000	.3238
15	.0000	.0000	.0000	.0000	.0000	.0000

CAESAR II Restraint Report Problem #1
Case 3 (EXP) D3 (EXP) = D1-D2

	Forces (lb.)			Moments (ft. lb.)			
Node	FX	FY	FZ	MX	MY	MZ	Type
5	-10934.	10934.	0.	0.	0.	-80360.	Rigid Anchor
15	10934.	-10934.	0.	0.	0.	80360.	Rigid Anchor

CAESAR II Force/Stress Report Problem #1
Case 3 (EXP) D3 (EXP) = D1-D2

Data	Forces (lb.)			Moments (ft. lb.)					(lb./sq. in.)	
Point	FX	FY	FZ	MX	MY	MZ	SIFI	SIFO	Code	Allow.
5	10933	-10933	0	0	0	80360	1.00	1.00	24455	28250
10	-10933	10933	0	0	0	37179	2.08	1.73	23482	28250
10	10933	-10933	0	0	0	-37179	1.00	1.00	11314	28250
15	-10933	10933	0	0	0	-80360	1.00	1.00	24455	28250

Figure 3.18 The “L” layout

CAESAR II (Ver. 2A) All Properties Listing (Pipe)
Job Description: B31.3 Sample #2

	From / To	X / Delta Y / Z	Dia. / Wall Thk. / Ins. Thk.	Temp 1 / Temp 2 / Temp 3	Pressure 1 / 2	Elastic Mod. / Poissons R. / Corosion	Pipe D / Insul. D / Fluid D			
Bend	5.	.000	10.750	750.00000	.0	.2790E+08	.2899	Bend Radius = 15.000	Fitting Thk. = .5000	
	10.	-6.000 ft.	.500	.00000	.0	.292000	.0000			
		.000	.000	.00000	Mat #1	.000000	.0000	RSTR Node = 5.	DIR = A	CN = 0.
								STIF=.100000E+13	GAP = .0000	MU = .00000
								Code B31.3	SC = 20000.00	
								SH1 = 13000.00	SH2 = .00	SH3 = .00
								F1 = 1.00	F2 = 1.00	F3 = 1.00
Bend	10.	.000	10.750	750.00000	.0	.2790E+08	.2899	Bend radius = 15.000	Fitting Thk. = .5000	
	15.	.000	.500	.00000	.0	.292000	.0000			
		-12.000 ft	.000	.00000		.000000	.0000			
Strt	15.	.000	10.750	750.00000	.0	.2790E+08	.2899	RSTR Node = 20.	DIR = A	CN = 0.
	20..	-6.000 ft	.500	.00000	.0	.292000	.0000	STIF =.100000E+13	GAP =.0000	MU = .00000
		.000	.000	.00000		.000000	.0000			

Figure 3.19 The "Z" shape layout.

CAESAR II Force/Stress Report Problem #2
Case 3 (EXP) D3(EXP)=D1-D2

Data	Forces (lb.)			Moments (ft. lb.)					(lb./sq. in.)	
Point	FX	FY	FZ	MX	MY	MZ	SIFI	SIFO	Code	Allow.
5	0	-10272	-33599	139966	0	0	1.00	1.00	42594	28250
10	0	10272	33599	48792	0	0	2.08	1.73	30818	28250
10	0	-10272	-33599	-48792	0	0	1.00	1.00	14848	28250
15	0	10272	33599	-19633	0	0	2.08	1.73	12400	28250
15	0	-10272	-33599	19633	0	0	1.00	1.00	5974	28250
20	0	10272	33599	139966	0	0	1.00	1.00	42594	28250

Pipe Supports

The purpose of pipe supports [¶321] is to control the weight effects of the piping system, as well as loads caused by pressure thrust, vibration, wind, earthquake, shock, and thermal displacement. The weight effects to be considered for support design shall be the greater of operating, (including thermal expansion loads) or hydrotest loads (unless provisions are made for temporary supports during hydrotest).

The B31.3 guidance for pipe support types and materials of construction is presented in B31.3 Table 326.1, Listed Standard, MSS SP-58. The material selection for clamps and bolts, for example, is of particular importance in elevated temperature service. SP-58 assists in the selection of a clamp material, for example in 750°F service. A review of the tables in SP-58 reveals that carbon steel clamp material would not be suitable, nor would the common type bolting, ASTM A 307, used in clamps. The designer is guided to use an alloy steel for the clamp such as ASTM A 240 and ASTM A 193 Grade B7 bolts.

Pipe support spacing is an important consideration in the design of piping systems. The permissible mid-span deflection concept, "y", is one technique commonly employed for support spacing. This technique is based on a specified mid-span "y" deflection of the supported pipe considering the pipe, contents, and insulation weights. The equation for determining the distance (L) between pipe supports is:

$$L = \left[\frac{yEI}{17.1W}\right]^{1/4}$$

where

L = pipe support spacing, feet
y = permissible mid-span deflection
E = modulus of elasticity at design temperature
I = moment of inertia of pipe = $\pi\frac{\left(D_o^4 - D_i^4\right)}{64}$, where "$D_o$" and "$D_i$" are the pipe OD and ID.
W = weight of supported pipe, including pipe, contents, insulation, lb/ft.

An example of the application of this mid-span deflection approach follows:

What is the span of a seamless ASTM A 106 Grade B, 6.625 inches OD, 0.28 inch wall thick, water-filled pipe with 3 inches of insulation with a design temperature of 400°F? The specifications limit the mid-span deflection to 0.25 in.

Solution:

Determine the uniform load, pounds per foot.
Pipe = 19.0 lbs/ft
Water = 12.5 lbs/ft
Insulation = 7.6 lbs/ft (85% magnesia calcium silicate)

then $W = 39.1$ lb/ft

$I = (\pi/64)(D_o^4 - D_i^4)$, $D_o = 6.625$, $D_i = 6.065$

$I = 28.14\ in^4$

$E = 27.7 \times 10^6$ psi, Table C-6, $C \leq 0.3$ at 400°F

$$\text{finally, } L = \left[\frac{0.25 \times 27.7 \times 10^6 \times 28.14}{17.1 \times 39.1}\right]^{1/4} = 23 \text{ feet span.}$$

The pipe support spacing of 23 feet will limit the mid span deflection to ¼ inch.

Variable spring supports are often necessary in piping systems where there is a difference between operating temperature and the installed temperature. For example, if resting supports were selected for a piping layout, and thermal expansions were to cause the pipe to lift off the support causing an increase in pipe load on adjacent supports or on load-sensitive equipment such as pumps or turbines, then a spring support may be required. This increase in load may be more than the adjacent support or equipment can safely accomodate. A variable spring may be the preferred pipe support for these instances where pipe lift-off would otherwise occur. The designer must decide which type of spring support to use and then size the spring. The procedure for sizing springs is demonstrated in the following example.

Assume a "type B" spring hanger support has been selected to carry an operating load of 1,300 pounds and the pipe will thermally translate 0.5 inch down at the support location (Figure 3.20). What spring size should be specified?

Table 3.10 can be used to size the spring. The objective is to select a spring size that will have a load-carrying range (considering the thermal translation), within the heavy horizontal lines shown near the top and bottom of the load table. The load for the spring to carry is to be the weight in the normal operating condition. For the conditions given (1,300 pounds), the table lists for this load a size 10 spring about half way between the heavy lines. The translation being 0.5 inch down, for our spring hanger, this translation would increase the load of the spring being that the style of spring selected is a spring hanger. This being the case, the cold load or the spring preset installed load must be less than the operating load. To find the cold load, move vertically in the table in the direction opposite to that of the pipe movement, the distance of the pipe translation (0.5 inch measured in the working range columns listed as figure 98, 268, or 82). The most common spring figure is the 268, a 0.5 inch translation up for this figure will produce a cold load of 1,170 pounds, (move vertically in the table in the direction opposite that of the pipe movement the distance of the pipe translation). The spring rate for the size 10 hanger is 260 pounds per inch. The % variability for this example is 10%, the spring is sized.

A test to determine if the correct spring size has been selected is to calculate the spring percent variability. The variability determined by this calculation should not exceed 25%; for springs near load-sensitive equipment, a spring with a lower variability should be selected (e.g., one in the 6 to 8% range). The equation for calculating the percent of variability is:

$$\%\ \text{Variability} = \frac{\text{Movement x Spring Rate}}{\text{Operating Load}} \text{x } 100$$

Table 3.10 Spring Size Selection Table

Grinnell

	Working Range*, in.					Hanger Size															Spring Deflection, in.		
	Quad.	Tri.	Fig. 98	Fig. 268	Fig. 82	000	00	0	1	2	3	4	5	6	7	8	9	10	11	12	Fig. 82	Fig. 268	Fig. 98
						7	19	43	63	81	105	141	189	252	336	450	600	780	1020	1350	0	0	0
						7	20	44	66	84	109	147	197	263	350	469	625	813	1063	1406			
						8	22	46	68	88	114	153	206	273	364	488	650	845	1105	1463			
						9	21	48	71	91	118	159	213	284	378	506	675	878	1148	1519			
Working Range	0	0	0	0	0	10	26	50	74	95	123	165	221	294	392	525	700	910	1190	1575	¼	½	1
						11	28	52	76	98	127	170	228	305	406	544	725	943	1233	1631			
						12	30	54	79	101	131	176	236	315	420	563	750	975	1275	1688			
						12	31	56	81	105	136	182	244	326	434	581	775	1008	1318	1744			
	2	1½	1	½	¼	14	34	58	84	108	140	188	252	336	448	600	800	1040	1360	1800	½	1	2
						14	35	59	87	111	144	194	260	337	462	619	825	1073	1403	1856			
						15	38	61	89	115	149	200	268	357	476	638	850	1105	1445	1913			
						16	40	63	92	118	153	206	276	368	490	656	875	1138	1488	1969			
	4	3	2	1	½	17	41	65	95	122	158	212	284	378	504	675	900	1170	1530	2025	¾	1½	3
						18	43	67	97	125	162	217	291	389	518	694	925	1203	1573	2081			
						19	45	69	100	128	166	223	299	399	532	713	950	1235	1615	2138			
						20	47	71	102	132	171	229	307	410	546	731	975	1268	1658	2194			
	6	4½	3	1½	¾	21	49	73	105	135	175	235	315	420	560	750	1000	1300	1700	2250	1	2	4
						21	50	74	108	138	179	241	323	431	574	769	1025	1333	1743	2306			
						22	53	76	110	142	184	247	331	441	588	788	1050	1365	1785	2363			
						23	55	78	113	145	188	253	339	452	602	806	1075	1398	1828	2419			
	8	6	4	2	1	24	56	80	116	149	193	258	347	462	616	825	1100	1430	1870	2475	1¼	1½	5
						25	58	82	118	152	197	264	354	473	630	844	1125	1463	1913	2531			
						26	60	84	121	155	201	270	362	483	644	863	1150	1495	1955	2588			
						27	62	86	123	159	206	276	370	494	658	881	1175	1528	1998	2644			
	10	7½	5	2½	1¼	28	64	88	126	162	210	282	378	504	672	900	1200	1560	2040	2700	1½	3	6
						28	66	89	129	165	214	288	386	515	688	919	1225	1593	2083	2756			
						29	68	91	131	169	219	294	394	525	700	938	1250	1625	2125	2813			
						30	70	93	134	172	223	300	402	536	714	956	1275	1658	2168	2869			
						31	72	95	137	176	228	306	410	546	728	975	1300	1690	2210	2925	1¾	3½	7
						Spring Rate - lb. per in.																	
					82			30	42	54	70	94	126	168	224	300	400	520	680	900	82		
					268	7	15	15	21	27	35	47	63	84	112	150	200	260	260	450	268		
					98			7	10	13	17	23	31	42	56	75	100	130	170	225	98		
					Triple			5	7	9	12	15	21	28	37	50	67	87	113	150			
					Quadruple			4	5	7	9	12	16	21	28	38	50	65	85	113			

*Figure numbers based on Grinnel spring hangers.

For the previous example:

$$\% \text{ Variability} = \frac{0.5 \text{ x } 260}{1300} \text{ x } 100 = 10\%$$

This spring size satisfies the variability requirement.

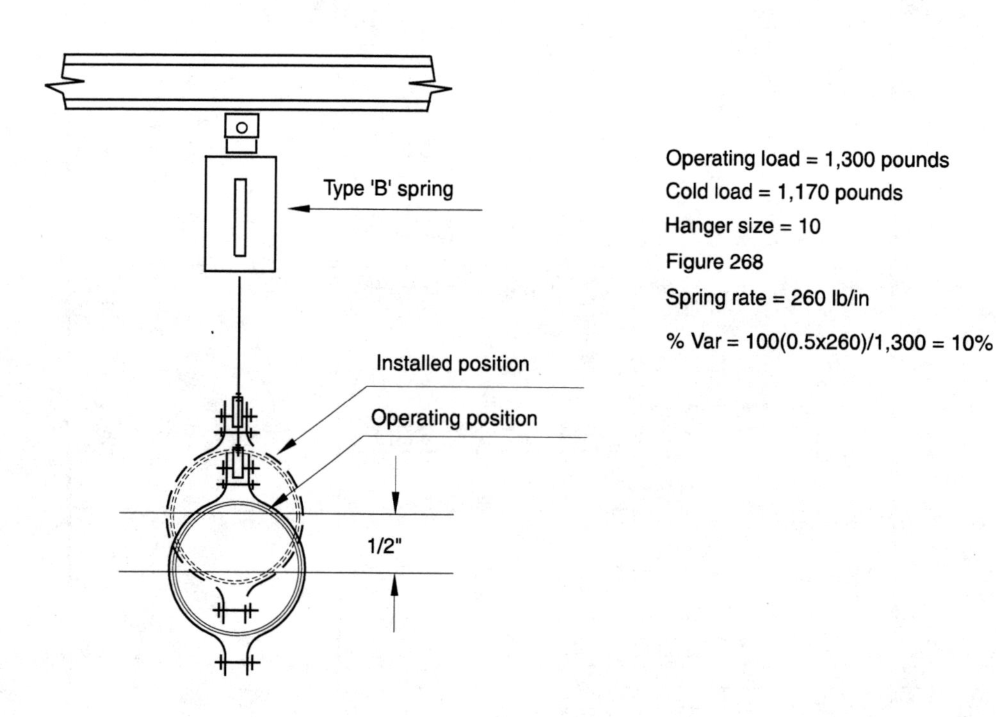

Figure 3.20 Spring hanger.

Chapter 4

LIMITATIONS ON PIPING AND COMPONENTS

The suitability of all pipe and components used in any system is the responsibility of the designer [¶300 (2)]. B31.3 provides guidance [¶305.2] for the designer to judge the suitability of pipe and components, particularly where limitations are known to exist. This guidance on limitations is provided relative to two operating regimes, these operating regimes are *fluid service categories* and *severe cyclic conditions*.

Fluid Service Categories

B31.3 recognizes the following three fluid service categories and a special design consideration based on pressure. It is the owners responsibility to specify the fluid service category and to determine if the high pressure requirement of B31.3 Chapter IX are applicable. These fluid categories and pressure concern are:

1. Normal Fluid Service (see B31.3 ¶300.2 "Fluid Service") (piping designed, fabricated, and inspected in accordance with the first seven chapters of B31.3)
2. Category D Fluid Service
3. Category M Fluid Service (Chapter 8, B31.3)
4. High Pressure Piping (B31.3, Chapter 9)

Category D Fluid Service is defined as having the following characteristics [¶300.2]:

- nonflammable
- nontoxic
- not damaging to human tissue
- the design gage pressure does not exceed 150 psig
- the design temperature is from -20°F to 366°F. 366°F is the saturation temperature of steam at 150 psig.

The pipe material limited to Category D Fluid Service is [¶305.2.1]:

- API 5L, Furnace Butt-Welded
- ASTM A 53, Type F
- ASTM A 134 made from other than ASTM A 285 plate

Components limited to use only in Category D Fluid Service are:

- miter bends that make a change in direction greater than 45° at a single joint [¶306.3.2]
- straight threaded couplings [¶314.2.1(d)]
- caulked joints [¶316]

Category M Fluid Service is defined [¶300.2] as a service in which a single exposure to a very small quantity of toxic fluid can produce serious irreversible harm to persons on breathing or bodily contact, even when prompt restorative measures are taken.
The *normal fluid service* is defined as all other fluid services that are not Category D or M, and is designed in accordance with the first seven chapters of the Code.

High pressure piping is a relatively new service covered in the Code. This coverage first appeared as Chapter IX of Addendum C of the B31.3 1984 Code Edition. High Pressure Piping service is defined as that in which the pressure is in excess of that allowed by the ASME B16.5 flange class ratings. If the owner designated a system as high pressure fluid service, then all the requirements of B31.3 Chapter IX shall be met.

The responsibility for categorizing fluid services lies with the plant owner. The Code assists the owner in categorizing category M fluid services in B31.3 Appendix M. If a fluid service is determined to be Category M, then the piping system shall conform to the first seven chapters of the Code as modified by Chapter VIII. A review of Chapter VIII reveals the modified requirements as well as additional pipe and component limitations.

Severe Cyclic Conditions

Severe cyclic conditions are those in which "S_E", the displacement stress range, exceeds 0.8 S_A, the allowable stress range, and the equivalent number of cycles exceeds 7000, or other conditions which the designer determines will produce an equivalent effect [¶300.2]. For piping systems determined to be severe cyclic, the designer must insure that the piping components selected for use in the system, are not prohibited from use by the B31.3 Code.

Often, the occurrence of severe cyclic conditions can be circumvented by piping layout changes, component selection, or other means while the piping is in the design phase. However, if severe cyclic conditions cannot be mitigated, the Code places many limitations on these systems as presented in Chapter II, Part 3. Also, piping operating under severe cyclic conditions must conform to more rigorous weld acceptance criteria than piping operating in the normal fluid service conditions. This topic is discussed later in the examination section.

Chapter

5

MATERIALS

Introduction

This chapter's discussion of materials and the B31.3 Code is based largely on the three premises listed below:

a) B31.3 assumes that users have some understanding of material classification systems, material specifications, and material properties. Experience shows, however, that the level of understanding among users varies widely and is often limited to a few grades of carbon steel or to a specific alloy system employed by the user on a regular basis. Consequently, this chapter begins with information on material classification systems and material specifications and is aimed at enhancing the user's overall understanding of materials.
b) B31.3 is a safety code focusing primarily on mechanical design, mechanical properties, and resulting pressure integrity. The Code lists a wide variety of materials which can be considered "prequalified" for use based on their inherent properties [¶323]. As part of the materials listing, the Code includes allowable stress values as a function of design temperature, and some helpful notes relating to material behavior under various service conditions [Tables A-1 and A-2, and Appendix F].
c) Although B31.3 does list acceptable materials and does provide certain prohibitions, limitations, conditions, and precautions on the use of acceptable materials, it does not prescribe which material to use for a specific application. Remember, the Code focus is on mechanical integrity, and it provides limited direction with respect to the suitability of any material for a particular process environment. Evaluation of expected material behavior for a given set of process conditions, including critical examination of the prohibitions, limitations, conditions, and precautions listed in the Code, generally requires input from a material's specialist.

Material Classification Systems and Specifications

The language of materials can be complicated. During discussions with material specialists, one often has the feeling of visiting a foreign country. Nevertheless, efficient use of B31.3 requires some basic comprehension of Code layout and materials technology, especially material classification systems and material specifications.

For example, consider locating an allowable stress for a particular austenitic stainless steel pipe material such as ASTM Standard A 312 Type 316L. An uninformed user could end up searching through the approximately 50 pages of information constituting B31.3 Table A-1. However, the search would be considerably less difficult if the user understood the following basic layout of Table A-1.

a) Table A-1 uses broad generic descriptors to classify listed materials under headings such as Carbon Steel, Nickel and Nickel Alloys, and Titanium and Titanium Alloys. These generic descriptors are typically located toward the top left and right sides of each page of the table, just below the table header.
b) Within each broad generic material descriptor, Table A-1 further classifies materials according to product form (e.g., plates and sheets, pipe and tube, forgings, etc.).
c) Finally, for each product form, Table A-1 lists materials according to nominal composition, material specification, and grade. Grades are described in terms of several standardized alphanumeric designation systems which depend upon the alloy system under consideration, and are described later in this chapter.

The preceding discussion of Table A-1 may appear simplistic, especially if you work with many different systems on a regular basis. However, there is a lot more to material identification, particularly since the manner of identification often depends upon the level and type of communication.

As another example, consider materials selection during the conceptual or front end engineering of a major project. At this stage of design, it is generally not prudent to specify piping materials using restrictive specification and grade designations such as ASTM A 312 Type 316L. Later in the project, one could easily ask, why can't we use Type 316, Type 304L, or Type 304? Alternatively, even though generic designations during front end design allow for flexibility in specific material grade selection during detailed mechanical design, there are situations where the only suitable material candidates are proprietary alloys offered by a select list of manufactures (e.g., Incoloy 825, Nicrofer 4221, Hastelloy C-22, etc.).

In summary, the three primary methods of identifying materials are:

a) generic designations,
b) trade names or proprietary designations, and
c) standardized alphanumeric designations.

These material designation systems are discussed in more detail under the next three headings.

Generic Description

Classification of materials by generic description involves the grouping of materials into broad categories according to certain attributes such as general composition, mechanical properties, product form, or end use. There are no precise rules governing which attributes to apply in defining material groups, and the level of detail afforded the classification system depends largely on the level of detail needed to communicate specific ideas. Consequently, materials may be generically grouped according to very broad characteristics, for example metal or nonmetal, ferrous or nonferrous, or cast or wrought. Alternatively, materials may be placed in more narrowly defined generic groups such as mild steel, 3XX series stainless steel, or NiCrMo alloy.

With piping materials, generic grouping based on alloy content is most popular. These groups usually reflect the primary alloy content, and may include varying levels of complexity depending upon the extent to which one needs to communicate specific material needs. Table 5.1 gives an indication of the progression from simple generic descriptors, to complex generic descriptors which may involve some elements of a standardized classification system (e.g., 300 series austenitic stainless steel).

Table 5.1 Levels of Generic Classification of Materials

Simple	Intermediate	Complex
Carbon Steel	Low Carbon Steel	Fully Killed, Low Carbon Steel
Low Alloy Steel	Cr-Mo Steel	2¼Cr-1Mo Steel
Stainless Steel	Austenitic Stainless Steel	300 Series Austenitic Stainless Steel
Nickel Alloy	High Nickel Alloy	NiCrMo Alloy

Generic material descriptions are frequently used during the early stages of a project, including project definition, conceptual design, front end design, preliminary design, process design, and/or budget estimation. For materials selection purposes during these stages, the user must be aware of Code requirements, but is not looking for a precise solution for each piping system. Rather, the user should be looking at more global issues including resistance of generic material groups to various forms of corrosion, material cost and availability for various product forms, delivery times, need for qualification testing, and existence of suitable forming and joining technology.

Trade Names and Proprietary Designations

Trade names are used by manufacturers to uniquely identify their materials and products. Sample trade names include Inconel 625, Incoloy 825, Hastelloy C-276, Carpenter 20Cb-3, Allegheny-Ludlum Al-6XN, Mather & Platt Xeron 100, Lincoln Fleetweld 5P+, and VDM 1925hMo.

Although there are definite commercial reasons for the existence of trade names (e.g., typically to induce purchasers to specify and buy only the product of a particular manufacturer), many manufacturers and trade associations publish trade name equivalency charts. Consequently, there is usually no need to restrict material selections through use of a single trade name. However, two exceptions do exist where it may be necessary to specify materials by trade name. The exceptions are:

a) materials of very recent development, which may still be protected by patent rights, and
b) sophisticated materials required for very severe service situations, where all potential manufacturers may not be equally capable of making the same quality of product. (For certain high alloy materials, minor chemistry or processing modifications can dramatically affect alloy performance.)

Standardized Alphanumeric Descriptors

Throughout the twentieth century, industry groups, government organizations, and trade associations all recognized the need for standardization of material designations. Numerous alphanumeric material designation systems were developed and continued to evolve nationally and internationally in a manner consistent with the introduction and development of new materials.

Because of the vast numbers of different materials in common use and the number of standardizing bodies involved in developing designation systems, naming conventions for materials can be overwhelming. For metallic materials, industry is gradually moving toward use of the Unified Numbering System (see 5.2.3.5 below) for material classification; however, other systems still prevail and probably will for many years to come. B31.3 identifies materials using a variety of designation systems, so it is necessary to understand the basis for several of the "standardized" material classification systems. Many of these are explained in the following paragraphs.

American Iron and Steel Institute (AISI)

The AISI numbering system for carbon and low alloy steels is essentially a four digit system which may have prefixes and suffixes attached to render more specific meaning. Table 5.2 shows a partial list of the main carbon and low alloy groups of the AISI system. In most cases, the user must refer to tables for exact composition ranges of AISI steels, but with some experience, the following two rules can be useful in identifying the more common grades.

a) The first and second digits of the alloy designation indicate the primary and secondary alloy classes to which the steel belongs.
b) The third and fourth digits (and the fifth for some groups) indicate the average carbon content in hundredths of weight percent.

For example, 4100 series steels are alloyed with chromium and molybdenum (nominally 1%Cr-0.2%Mo, with some exceptions), so a 4140 steel would be a 1%Cr-0.2%Mo steel with about 0.40%C.

Table 5.2 AISI Designation System for Carbon and Low Alloy Steels

Carbon Steels	
10xx	Plain Carbon Steel
11xx	Plain Carbon Steel, Resulfurized
1200	Plain Carbon Steel, Resulfurized and Rephosphorized
Low Alloy Steels	
13xx	1¾% Manganese (1.60-1.90%) steels
2100	1% Nickel (1%)
23xx	3½% Nickel (3.25-3.75%) steels
25xx	5% Nickel (4.75-5.25%) steels
31xx	1¼% Nickel (1.10-1.40%) - Chromium (0.55-0.75 or 0.70-0.90%) steels
33xx	3½% Nickel (3.25-3.75%) - Chromium (1.40-1.75%) steels
40xx	¼% Molybdenum (0.15-0.25 or 0.20-0.30%) steels
41xx	1% Chromium (0.40-0.60, 0.70-0.90, or 0.80-1.I0%) - Molybdenum (0.08-0.15, 0.15-0.25%, or 0.25-0.35) steels
43xx	1¾% Nickel (1.65-2.0%) - Chromium (0.40-0.60 or 0.70-0.90%) - Molybdenum (0.20-0.30%) steels
46xx	1¾% Nickel (0.70-1.00, 1.40-1.75, 1.65-2.00%) - Molybdenum (0.15-0.25, 0.20-0.30%) steels
48xx	Nickel (3.25-3.75%) - Molybdenum (0.20-0.30%) steels
50xx	Chromium (0.20-0.35 or 0.55-0.75%) steels
51xx	Chromium (0.80, 0.90 or 1.05%) steels
5xxxx	Chromium steel (0.50, 1.00, or 1.45%) with carbon (0.95-1.10%)
61xx	Chromium (0.80 or 0.95%) - Vanadium (0.10 or 0.15% min.) steel
86xx	Nickel (0.40-0.70%) - Chromium (0.40-0.60%) - Molybdenum (0.15-0.25%) steels
87xx	Nickel (0.40-0.70%) - Chromium (0.40-0.60%) - Molybdenum (0.20-0.30%) steels
92xx	Manganese (0.85%) - Silicon (1.8-2.2%) steels
93xx	Nickel (3.0-3.50%) - Chromium (1.0-1.4%) - Molybdenum (0.08-0.15%) steels
94xx	Manganese (1.0%) - Nickel (0.30-0.60%) - Chromium (0.3-0.5%) - Molybdenum (0.08-0.15%) steels
97xx	Nickel (0.40-0.70%) - Chromium (0. 1-0.25%) - Molybdenum (0. 15-0.25%) steels
98xx	Nickel (0.85-1.15%) - Chromium (0.70-0.90%) - Molybdenum (0.20-0.30%) steels
99xx	Nickel (1.00-1.30%) - Chromium (0.40-0.60%) - Molybdenum (0.20-0.30%) steels
Note: A "B" inserted between the second and third letter denotes a Boron alloyed steel	

Prefix and suffix letters may be added to the AISI numbers shown in Table 5.2, but it is very common to see the alloy number without the prefix, since composition is usually the main item of concern. Nevertheless, for guidance, some possible prefix and suffix letters are listed in Table 5.3.

Table 5.3 AISI Prefix and Suffix Meanings

Prefix		Suffix	
A	Basic Open-Hearth Alloy Steel	**A**	Restricted Chemical Composition
B	Acid Bessemer Carbon Steel	**B**	Bearing Steel Quality
C	Basic Open-Hearth Carbon Steel	**C**	Guaranteed Segregation Limits
D	Acid Open-Hearth Carbon Steel	**D**	Specified Discard
E	Basic Electric Furnace Steel	**E**	Macro-Etch Tests
TS	Tentative Standard Steel	**F**	Rifle Barrel Quality
Q	Forging Quality, or Special	**G**	Limited Austenitic Grain Size
R	Re-rolling Quality Billets	**H**	Guaranteed Hardenability
		I	Non-Metallic Inclusions Requirements
		J	Fracture Test
		T	Extensometer Test
		V	Aircraft Quality or Magnaflux Testing

American Iron and Steel Institute (AISI) System for Intermediate and High Alloy (Stainless) Steels

AISI also has a three digit system for designating intermediate and high alloy steels. For the most part, these are stainless steels which by definition contain more than 11.5% chromium (i.e., 2XX, 3XX, 4XX, and 6XX series). Some members of the classification system are, however, intermediate alloy steels (i.e., 5XX series). Table 5.4 identifies the major groups and their common characteristics. Examples of AISI identifications include the 304, 316L, 410, and 630 high alloy stainless steels, and 502 intermediate alloy steel.

Table 5.4 AISI Designations for Intermediate and High Alloy (Stainless) Steels

Series No.	Group Characteristics
2xx	Chromium-Manganese-Nickel stainless steels; austenitic structure, nonmagnetic, not hardenable by quenching.
3xx	Chromium-Nickel stainless steels; austenitic structure, nonmagnetic, not hardenable by quenching.
4xx	Chromium stainless steels; martensitic or ferritic structure, magnetic, martensitic stainless steels can be hardened by quenching.
5xx	Straight chromium steels; ferritic structure, magnetic, can be hardened by quenching.
6xx	Stainless steels which are precipitation hardenable through incorporation of small additions of Al, Ti, Cu, or other alloying elements to permit strengthening by aging heat treatments; structures are variable and may be mixed.

Alloy Castings Institute (ACI)

Cast stainless steels are often specified on the basis of composition, using a designation system established by the Alloy Casting Institute (ACI). ACI designations such as CF3 and CF8M, have been adopted by many standards issuing organizations such as ASTM and are included in B31.3 for stainless steel castings. Even though the wrought AISI stainless steel designations (e.g., 316) are often applied to castings, the ACI designations are preferred since the composition of the cast material will not be identical to the composition of the "equivalent" wrought material.

An explanation of the ACI designation system, based on the common alloys CF8M and HK40, follows.

a) The first letter of the cast stainless steel designation system identifies the intended service application of the alloy. The letter C indicates corrosion-resistant service (as in CF8M), and the letter H indicates heat-resistant service (as in HK40).
b) The second letter indicates the approximate location of the alloy on the iron-chromium-nickel (FeCrNi) ternary diagram. For users familiar with the diagram, the second letter does provide an indication of the nominal iron, nickel, and chromium content, but most people would have to obtain alloying information from a material specification.
c) The third and fourth digits indicate carbon content of the alloy. For corrosive service (C) castings, the third and fourth digits represent the maximum permitted carbon content in units of 0.01% (e.g., CF8M has a maximum of 0.08% carbon). For heat resistant (H) castings, the third and fourth digits represent the midpoint of the carbon range in units of 0.01% with a ±0.05% limit (e.g., HK40 contains nominally 0.40% carbon, ±0.05%).
d) Additional letters following the numerals representing carbon content refer to special chemical elements added to the alloy, and may include M for molybdenum, C for columbium, Cu for copper, and W for tungsten (e.g., CF8M contains 2.0 - 3.0% molybdenum). There are two exceptions to this rule: the letter A indicates "Controlled Ferrite," and the letter F indicates "Free Machining."

Aluminum Association (AA)

The designation of aluminum and aluminum alloys is complex and is described in detail by ANSI H35.1. The designation system integrates three items of information:

a) product form (wrought products, cast products, and ingots),
b) chemical composition (by a method of alloy grouping), and
c) temper.

Table 5.5 outlines the principle aluminum alloy designations based on product form and nominal composition of each group. The temper component of the designation system is described later.

In the designation system, distinction with respect to product form is quite simple.

a) AA numbers without a decimal between the third and fourth digits are wrought products. These are listed in the two left columns of Table 5.5.

b) AA numbers with a decimal between the third and fourth digits (i.e., two right columns of Table 5.5) are:

   i) cast products if the digit after the decimal is a 0; and

   ii) ingot products if the digit after the decimal is a 1 or a 2, which differentiate ranges of ingot composition.

Table 5.5 Aluminum Association (AA) Wrought and Cast Alloy Designations

Wrought Alloy System		**Cast Alloy System**	
1xxx	Aluminum, >99.00%	1xx.x	Aluminum, >99.00%
2xxx	Copper	2xx.x	Copper
3xxx	Manganese	3xx.x	Silicon, Copper, and/or Magnesium.
4xxx	Silicon	4xx.x	Silicon
5xxx	Magnesium	5xx.x	Magnesium
6xxx	Magnesium and Silicon	6xx.x	Unused series
7xxx	Zinc	7xx.x	Zinc
8xxx	Special Alloys	8xx.x	Tin
9xxx	Unused Series	9xx.x	Other Elements

The chemical compositions of wrought aluminum and aluminum alloys are described by four-digit numbers with the following meaning:

a) The first digit indicates the primary alloy group as listed in Table 5.5.

b) The second digit indicates modifications of the original alloy or impurity limits.

c) The last two digits identify the specific aluminum alloy or indicate the aluminum purity.

The chemical compositions of cast aluminum and aluminum alloys are described by the three numbers preceding the decimal, as follows:

a) The first digit indicates the primary alloy group as listed in Table 5.5.

b) The second and third digits identify the aluminum alloy or indicate the aluminum purity.

Recall that the last digit, which appears after the decimal point, indicates product form (i.e., castings or ingots).

In addition to the product and chemistry component of the designation system, there is a temper component which exists for all forms of wrought and cast for aluminum and aluminum alloys, except ingots. The temper component is based on the sequence of treatments used to make a product. It follows the alloy designation and is separated from the four digit alloy designation by a hyphen (e.g., ASTM B 241 Gr. 5083-O, ASTM B 210 Gr. 1060-H113, ASTM B 241 Gr. 6061-T6). Basic temper designations consist of the letters shown in Table 5.6.

Table 5.6 Basic Temper Designations for Aluminum and Aluminum Alloys

Temper	Meaning	Application
F	As Fabricated	Products of shaping processes in which no special control over thermal conditions or strain-hardening is employed. For wrought products, there are no mechanical property limits.
O	Annealed	Wrought products which are annealed to obtain the lowest strength temper and to cast products which are annealed to improve ductility and dimensional stability. The O may be followed by a digit other than zero.
H	Strain-Hardened	Wrought products which have their strength increased by strain hardening, with or without supplementary thermal treatments to produce some reduction in strength. The H is always followed by two or more digits.
W	Heat-Treated	An unstable temper applicable only to alloys which spontaneously age at room temperature after solution heat-treatment. This designation is specific only when the period of natural aging is indicated.
T	Thermally Treated	Products which are thermally treated, with or without supplementary strain-hardening, to produce stable tempers other than F, O, or H. The T is always followed by one or more digits.

Subdivisions of the H and T basic tempers are indicated by one or more digits following the letter. These digits indicate a specific sequence of basic treatments, recognized as significantly influencing the characteristics of the product. For subdivisions of the H temper, the first digit following the H indicates a specific combination of basic operations which influence characteristics such as hardness and strength, as shown by Table 5.7. For subdivisions of the T temper, the numerals 1 through 10 following the T indicate a specific sequence of basic thermal treatments which influence characteristics such as hardness and strength, as shown by Table 5.8.

Table 5.7 Subdivisions of the H Temper Designation

Temper	Meaning	Application
H1	Strain-Hardened Only	Products which are strain-hardened to obtain the desired strength without supplementary thermal treatment. The number following this designation indicates the degree of strain-hardening.
H2	Strain-Hardened and Partially Annealed	Products which are strain-hardened more than the desired final amount and then reduced in strength to the desired level by partial annealing. The number following this designation indicates the degree of strain-hardening remaining after the product has been partially annealed.
H3	Strain Hardened and Stabilized	Products which are strain hardened and whose mechanical properties are stabilized by a low temperature thermal treatment. The number following this designation indicates the degree of strain-hardening before the stabilization treatment.

Table 5.8 Subdivisions of the T Temper Designation

Temper	Meaning	Application
T1	Cooled from Elevated Temperature and then Naturally Aged	Products which are not cold worked after cooling from an elevated temperature shaping process, or in which the effect of cold work in flattening or straightening may not be recognized in mechanical property limits.
T2	Cooled from Elevated Temperature, Cold Worked, and then Naturally Aged	Products which are cold worked to improve strength after cooling from an elevated temperature shaping process, or in which the effect of cold work in flattening or straightening is recognized in mechanical property limits.
T3	Solution Heat Treated, Cold Worked, and then Naturally Aged	Products which are cold worked to improve strength after solution heat treatment, or in which the effect of cold work in flattening or straightening is recognized in mechanical property limits.
T4	Solution Heat Treated and then Naturally Aged	Products which are not cold worked after solution heat-treatment, or in which the effect of cold work in flattening or straightening may not be recognized in mechanical property limits.
T5	Cooled from Elevated Temperature and then Artificially Aged	Products which are not cold worked after cooling from an elevated temperature shaping process, or in which the effect of cold work in flattening or straightening may not be recognized in mechanical property limits.
T6	Solution Heat Treat and then Artificially Aged	Products which are not cold worked after solution heat treatment, or in which the effect of cold work in flattening or straightening may not be recognized in mechanical property limits.
T7	Solution Heat Treated and then Stabilized	Products which are stabilized after solution heat-treatment to carry them beyond the point of maximum strength to provide control of some special characteristic.
T8	Solution Heat Treated, Cold Worked, and then Artificially Aged	Products which are cold worked to improve strength, or in which the effect of cold work in flattening or straightening is recognized in mechanical property limits.
T9	Solution Heat Treated, Artificially Aged, and then Cold Worked	Products which are cold worked to improve strength.
T10	Cooled from Elevated Temperature, Cold Worked, and then Artificially Aged	Products which are cold worked to improve strength, or in which the effect of cold work in flattening or straightening is recognized in mechanical property limits.

If a variation to the sequence of basic operations applied to an alloy results in different characteristics, then additional digits may be added to the temper designation. For the H temper, one or two digits following the basic temper designation (i.e., H1, H2, or H3) indicate the degree of strain-hardening. The numeral 8 has been assigned to indicate tempers having an ultimate tensile strength equivalent to that achieved by a cold reduction of approximately 75 percent following a full anneal (e.g., ASTM B 210 Gr. 3003-H18). Tempers between O (annealed) and 8 are designated by numerals 1 through 7. Material having an ultimate tensile strength about midway between that of the O temper and the 8 temper is designated by the numeral 4 (e.g., ASTM B 210 Gr. 1060 H14); about midway between O and 4 tempers by the numeral 2; and about midway between the 4 and 8 tempers by the numeral 6. The numeral 9 designates tempers whose minimum ultimate tensile strength exceeds that of the 8 temper by 2.0 ksi or more. For two-digit H tempers whose second digit is odd, the standard limits for ultimate tensile strength are exactly midway between those of the adjacent two digit H tempers whose second digits are even. A third digit following the designation H1, H2, or H3 may be used to indicate a variation of a two-digit temper (e.g., ASTM B 241 Gr. 1060 H112). It is used when the degree of control of temper or the mechanical properties are different from but close to those for the two-digit H temper designation to which it is added, or when some other characteristic is significantly affected.

For thermally treated aluminum alloys, additional digits, the first of which may not be zero, may be added to designations T1 through T10 to indicate a variation in treatment which significantly alters the characteristics of the product. Other variations of temper do exist, and it is generally necessary to consult ANSI H45.1 or ASTM standards for correct interpretation.

Unified Numbering System

The Unified Numbering System (UNS) provides a consistent means of identifying many of the alloy designation systems currently administered by societies, trade associations, and individual users and producers of metals and alloys. The UNS thereby avoids confusion caused by use of more than one identification number for the same metal or alloy, and by the opposite situation of having the same number assigned to two or more different metals or alloys. It also provides the uniformity necessary for efficient indexing, record keeping, data storage and retrieval, and cross referencing.

Like other designation systems discussed earlier in this chapter, the UNS is not a specification, since it establishes no requirements relative to product form, condition, mechanical properties, or quality. It is simply an identification system for metals or alloys for which controlling composition limits have been established in specifications published elsewhere. The use of the UNS is increasing rapidly, and many UNS references are now found in B31.3 (e.g., ASTM B 165 UNS No. N04400, ASTM B 467 UNS No. C71500).

The UNS has established eighteen series of numbers for metals and alloys, as shown in Table 5.9. Each UNS number consists of a single letter prefix followed by five digits. In many cases the letter suggests the family of metals identified; for example, A for aluminum, P for precious metals, and S for stainless steels.

The six character alphanumeric UNS number is a compromise between those who think that identification numbers should indicate as many characteristics of the material as possible, and those who believe that numbers should be short and uncomplicated for wide acceptance and use. Within certain UNS groups, some digits have special meaning, but each group is independent of the others in such significance. For example, where feasible, existing material designation systems were incorporated into the UNS. A carbon steel identified as AISI 1020 is designated by G10200 in the UNS. A free cutting brass identified by the Copper Development Association as C36000 is UNS C36000. A stainless steel identified as AISI 316L is also UNS S31603.

Table 5.9 Unified Numbering System (UNS) Designations

Ferrous Metals and Alloys		Nonferrous Metals and Alloys	
D00001-D99999	Specified mechanical properties steels	A00001-A99999	Aluminum and aluminum alloys
F00001-F99999	Cast irons and cast steels	C00001-C99999	Copper and copper alloys
G00001-G99999	AISI and SAE carbon and alloy steels	E00001-E99999	Rare earth and rare earth-like metals and alloys
H00001-H99999	AISI H-steels	L00001-L99999	Low melting metals and alloys
J00001-J99999	Cast steels (except tool steels)	M00001-M99999	Miscellaneous nonferrous metals and alloys
K00001-K99999	Miscellaneous steels and ferrous alloys	N00001-N99999	Nickel and nickel alloys
S00001-S99999	Heat and corrosion resistant (stainless) steels	P00001-P99999	Precious metals and alloys
T00001-T99999	Tool steels	R00001-R99999	Reactive and refractory metals and alloys
		Z00001-Z99999	Zinc and zinc alloys
Specialized Metals And Alloys			
W00001-W99999	Welding filler metals, covered and tubular electrodes, classified by weld deposit composition		

Identification of welding filler metals by UNS numbers has not caught on for general industry use, perhaps because the existing AWS classification system works well for most welding consumables.

For additional details on the Unified Numbering System, consult:

a) ASTM Standard E 527, "Standard Practice for Numbering Metals And Alloys" (or SAE Standard J1086).
b) The joint ASTM/SAE publication titled "Metals & Alloys in the Unified Numbering System."
c) Volume 00.01 of the ASTM Book of Standards (the index), which contains an extensive array of UNS materials listed under the heading "UNS..."
d) *The Metals Black Book*™ - Ferrous Metals published by CASTI Publishing Inc.

Common ASTM Carbon Steel Piping Materials

For everyday work, most piping systems are constructed from carbon steel. Material designations are seemingly inconsistent and random and, for the most part, knowledge of specifications and grades can only be gained with experience. Nevertheless, for practical guidance, specification and grade designations can be grouped according to product form and notch toughness properties, as in Table 5.10.

Table 5.10 Common ASTM Carbon Steel Piping Material Specifications and Grades

Product Forms	See Note(s)	ASTM Materials Without Impact Tests	ASTM Materials With Impact Tests
Pipe	2	A 53 Gr. B A 106 Gr. B	A 333 Gr. 1 A 333 Gr. 6
Flanges & Forged Fittings	3	A 105	A 350 Gr. LF2
Wrought Fittings		A 234 Gr. WPB	A 420 Gr. WPL6
Castings		A 216 Gr. WCB A 216 Gr. WCC	A 352 Gr. LCB A 352 Gr. LCC
Bolts, Studs, And Cap Screws	4,5	A 193 Gr. B7 A 193 Gr. B7M	A 320 Gr. L7 A 320 Gr. L7M
Nuts	4,5	A 194 Gr. 2H A 194 Gr. 2HM	A 194 Gr. 7 A 194 Gr. 7M

<u>Notes to Table 5.10</u>

(1) Column headings refer to impact testing requirements of the material manufacturing specifications.

(2) For ASTM A 53, the type of pipe should also be specified, where S=Seamless, E=Electric Resistance Welded, and F=Furnace Butt Welded. Note: Type S is normally used for pressure piping, Type E is sometimes used, but Type F is rarely used.

(3) Forged fittings include weldolets, threadolets, and sockolets (WOL, TOL, SOL).

(4) B7M, 2HM, L7M, and 7M refers to grades with strength and hardness control, typically for resistance to sulfide stress cracking in sour (H_2S) environments. See also NACE MR0175.

(5) Quenched and tempered low alloy bolting has excellent notch toughness at lower temperatures. B31.3 permits use of low alloy steel bolting to lower temperatures than would be expected for corresponding carbon steel piping components. See the "Min. Temp." column of B31.3 Table A-2 for lower limit of temperature.

Material Requirements of B31.3

Materials considerations are specifically covered in B31.3 Chapter III, but there are also material references in many other chapters. In fact, after stating the obvious in the first sentence of Chapter III (that "limitations and required qualifications for materials are base on their inherent properties") [¶323], B31.3 continues by referring back to ¶300(d) in Chapter 1. Such is the nature of the Code.

Fluid Service Categories and Materials

¶300(d) begins by setting out a global materials philosophy for B31.3. It refers to fluid service categories defined by the Code (see Table 5.11) and indicates that they affect selection and application of materials, components, and joints. So, within the context of the Code, the category of fluid service constitutes one of the issues to be considered during selection of materials, components, and joints by virtue of certain prohibitions, limitations, and conditions found scattered throughout the Code, including ¶323.4, ¶323.5, ¶F323, and Notes to Tables A-1 and A-2.

Table 5.11 Fluid Service Categories

Normal Fluid Service	Pertains to most piping covered by the Code and includes piping not classified within the other fluid services listed below [¶300.2].
Category D Service	Service in which the fluid is nonflammable, nontoxic, and not damaging to human tissue; the design pressure does not exceed 150 psig (1030 kPag); and the design temperature is from -20°F (-29°C) to 366°F (186°C) [¶300.2].
Category M Service	Service in which a single exposure to a very small quantity of toxic fluid can produce serious irreversible harm on breathing or body contact, even when prompt restorative measures are taken [¶300.2].
High Pressure (K) Service	Service which applies when designated by the owner, typically for pressures in excess of that allowed by ASME B16.5 Class 2500 rating, for the specified design temperature and material group [¶K300(a)].

Notes to Table 5-11

a) For exact descriptions and classification of fluid service, refer to B31.3.

b) *Severe cyclic conditions* are defined by B31.3 as "conditions which are vibrating, cyclic or fatigue in nature, with S_E exceeding $0.8S_A$ and with the equivalent number of cycles exceeding 7000; or other conditions which the designer determines will produce an equivalent effect [¶300.2]." Even though it may appear that severe cyclic conditions satisfy the B31.3 definition for a "fluid service," they have not been included as a separate fluid service in the above table because, when severe cyclic conditions exist, they are typically a subset of one of the other four listed fluid services.

Materials and Specifications [¶323.1]

B31.3 classifies materials as listed, unlisted, unknown, or reclaimed, and places conditions on the use of such materials. Table 5.12 summarizes the characteristics of each material classification.

In most cases, Code users deal with listed materials. These may be considered as materials which are "prequalified" for Code use based on inherent properties [¶323] and listed in B31.3 Tables A-1 and A-2. For pressure design purposes, the Code provides stress values for these listed materials as a function of temperature (since mechanical behavior is temperature dependent). However, the suitability of a particular material for a particular fluid service is beyond the scope of the Code [¶300(c)(6)]. A materials specialist should be consulted to ensure correct materials selection for a fluid service.

Temperature Limitations [¶323.2]

B31.3 recognizes that material properties and behavior in service are temperature dependent. A significant portion of B31.3 Chapter III deals with temperature limitations for materials, in particular lower temperature limits where impact testing may apply. The Code also imposes cautionary and restrictive temperature limits in Tables A-1 and A-2, and requires designers to verify that materials are suitable for service throughout the operating temperature range [¶323.2].

Table 5.12 Material Classifications [¶323.1]

Listed Materials	[¶323.1.1] Those materials or components which conform to a specification listed in B31.3 Appendix A, Appendix B, or Appendix K, or to a standard listed in B31.3 Table 326.1, A326.1 or K326.1. For pressure design, listed materials and allowable stress values are shown in B31.3 Table A-1. Because allowable stress values are provided for listed materials, these are most convenient to use.
Unlisted Materials	[¶323.1.2] Those materials which conform to a published specification covering chemistry, physical, and mechanical properties, method and process of manufacture, heat treatment, and quality control, and which otherwise meet the requirements of the Code.
Unknown Materials	[¶323.1.3] Those materials of unknown specification, not to be used for pressure containing piping components.
Reclaimed Materials	[¶323.1.4] Used materials which have been salvaged and properly identified as conforming to a listed or published specification.

Upper Temperature Limits [¶323.2.1]

Upper temperature limits for listed materials are the maximum temperatures for which a stress value or rating is shown directly in or referenced by the Code. The Code may also provide notes to the stress value tables, precautionary information in Appendix F, and/or restrictions within the text of Code. For example, from Table A-1, the upper temperature limit for ASTM A 106 Grade B pipe is 1100°F, even though there are two notes pertaining to use of the material above 800°F and 900°F, respectively. ¶F323.4(a)(2) and ¶F323.4(a)(4) also discuss these notes.

Of course, the Code does permit use of listed materials at temperatures above the maximum indicated by the stress value or rating, provided there is no prohibition in the Code [¶323.2.1(a)] and provided the designer verifies the serviceability of the material [¶323.2.1(b)]. Verification would typically involve material specialists with an engineering background and a "sound scientific program carried out in accordance with recognized technology" [¶323.2.4].

Lower Temperature Limits and Impact Testing [¶323.2.2]

Lower temperature limits for materials are established as a means of controlling risk of brittle fracture. Terms frequently used in lower temperature limit discussions include notch sensitivity, impact testing, Charpy testing, and notch brittleness.

For most Code users, the basic question to be answered is: "Do I need to use impact tested materials?" Answering the question can be complex and convoluted; however, the basic steps to determining the answer are listed below and are discussed in detail in the following paragraphs.

a) Select the design minimum temperature for the piping. This may involve process engineering and/or heat transfer specialists, and consideration of ambient temperature effects.

b) Obtain the minimum permissible temperature for the proposed piping materials according to B31.3 rules.
c) Follow the instructions of B31.3 to determine whether impact tests are required (e.g., Table 323.2.2).
d) If impact tests are required, consult the additional requirements of B31.3 regarding impact test methods and acceptance criteria.

Selecting the Design Minimum Temperature (DMT)

The first step in assessing the need for impact tested materials is to select a *design minimum temperature* [¶301.3.1]. Although the Code does describe many factors which should be considered in selecting the design minimum temperature, it does not prescribe exactly how these factors must be considered. Such decisions are usually passed on to other professionals (e.g., process, mechanical, or heat transfer) who must determine the lowest metal temperature which can be achieved during steady state (normal) and non-steady state (upset) operations, including start-up, planned shut-down, emergency shut-down, and depressuring (e.g., Joule Thompson (JT) cooling).

Regarding low ambient temperatures, significant debate can occur about their effect on the selection of design minimum temperature and, consequently, on the need for "expensive" impact tested materials. Companies working in cold temperature climates should have a policy to deal with this issue. In the absence of a policy, Table 5.14 (not a Code requirement) could be used for guidance in evaluating ambient temperature effects. The table considers typical winter conditions in western Canada and reflects a general quality philosophy regarding impact tested carbon steel materials. Specifically, impact tested carbon steels which satisfy only the minimum requirements of ASTM, ASME, and API standards do not guarantee resistance to brittle fracture, but the inherently fine grain size of these so called "low temperature" steels does provide some proven measure of improved resistance to brittle fracture initiation. Impact testing which satisfies B31.3 at or below design minimum temperature is generally accepted as evidence of sufficient resistance to brittle fracture for most common designs. However, several owners do impose supplementary toughness requirements.

Minimum Permissible Temperature for a Material

To obtain the minimum permitted temperature for a listed material, refer to the column heading "Min. Temp. °F" in B31.3 Table A-1. For each material, the column entry will be either the minimum permissible temperature (for most materials) or a letter code (for certain common carbon steel materials). If the entry is a letter code, one must refer to Fig. 323.2.2 to obtain the minimum permissible temperature for use of carbon steels without impact testing. (Note that the curve and letter system for carbon steels was introduced to B31.3 with the B31.3b-1994 Addenda. While this approach to minimum permissible temperature is similar to that of ASME Section VIII, Division 1 for Unfired Carbon Steels (UCS-66), and also similar to some owner specifications, expect several revisions to the B31.3 rules, as happened with introduction of these curves for carbon steel pressure vessels.

Table 5.14 Guide to Selection of Design Minimum Temperature (DMT)
Based on Location and Ambient Temperature

Piping Location	Location Comments	DMT °C	DMT °F
Outside	Piping exposed directly to ambient winter conditions	-46	-50
Unheated Enclosure	Piping contained within an unheated enclosure, where process heat will maintain the enclosure temperature above design minimum temperature.	-29	-20
Buried	Piping buried below the frost line.	-5	+23
Heated Enclosure	Piping contained within a heated enclosure.	0	+32

Bases of Development for Table 5.14

(1) The basis for use of -46°C (-50°F) as DMT for outside exposure is that most "low temperature" carbon steel material standards require impact testing at -46°C (-50°F). This temperature is also typical of the coldest winter temperatures for the provinces of western Canada.

(2) Note that ASTM A 20 impact test temperatures for plate depend upon thickness. Many manufacturers are aware of this dependence; however, it may be advisable to consider this issue during design and pre-award meetings. In some situations, it may be possible to consider an alternative DMT which would allow use of non-impact tested plate or plate impact tested at a higher temperature (e.g., -40°C (-40°F)).

(3) For unheated enclosures, the availability of process heat and the absence of direct wind chill removing process heat is considered justification for selection of a DMT above that used for direct outdoor exposure. The -29°C (-20°F) DMT is based on ASME B31.3 requirements prior to the 1993(b) addenda, which permitted most common carbon steel piping materials to be used down to -29°C (-20°F) without impact testing. With the current edition of ASME B31.3, users must now refer to Table A-1 and Figure 323.2.2 of the Code, to determine minimum thickness and temperature threshold for impact testing.

(4) Although unheated enclosures may experience occasional excursions below -29°C (-20°F) due to seasonal or process variations, process heat will normally keep piping and equipment warm. For extended excursions below -29°C (-20°F) (e.g., a winter shut-down/start-up), temporary heat or warm start-up procedures may be necessary and should be planned for during design.

(5) For buried piping, -5°C (+23°F) was chosen as a commonly applied temperature for prairie and southern foothills pipeline design. For some northern locations and large diameter pipelines, -15°C (+5°F) has been used.

(6) For heated enclosures, the freezing point of water (0°C (+32°F)) was chosen as the DMT on the basis that permanent heat is usually required for operation of the process and/or utilities (e.g., to prevent freezing of cooling water or drains). The 0°C (+32°F) DMT for heated enclosures may be convenient to avoid impact testing of some pressure vessels and thick pressure piping. A DMT higher than 0°C (+32°F) could be selected if warranted by the quality of enclosure, process variables, or testing and service requirements of the piping or vessel.

(7) This table does not apply to situations where the need for low temperature impact tested materials is dependent on process design (e.g., Joule Thompson effects on blowdown).

(8) For some designs, avoidance of low temperature impact tested materials may be justified on the basis of insulation and heat tracing, warm start-up procedures, temporary heat, or low stress as implied by B31.3 Table 323.2.2 Note 3.

As an example in the determination of minimum permissible temperature of a material, first consider the case of 6 NPS Sch. 40 (0.280 in. WT) ASTM A 312 Type 316L seamless pipe. From Table A-1 the minimum permissible temperature obtained from the "Min. Temp. °F" column is -325°F. As a second example, consider a piece of 6 NPS Sch. 160 (0.719 in. WT), ASTM A 106 Grade

B pipe. Table A-1 gives this material a letter code "B." Now, referring to Figure 323.2.2, for a thickness of 0.719 in. the minimum permissible temperature for use without impact testing is about +15°F.

Table 323.2.2 - Requirements for Low Temperature Toughness Tests

When the minimum permissible temperature for a listed material has been determined, refer to B31.3 Table 323.2.2 to determine requirements for low temperature toughness tests for base metal, weld metal, and heat affected zones. For a given type of material, impact testing requirements are provided for two situations.

a) Column "A" of the table applies to situations where the design minimum temperature is at or above the minimum permissible temperature for the proposed material, as given by Table A-1 or Figure 323.2.2. For most materials, note that Column "A" is actually divided into two columns: "(a) Base Metal" and "(b) Weld Metal and Heat Affected Zones."
b) Column "B" applies to situations where the design minimum temperature is below the minimum permissible temperature for the proposed material, as shown in Table A-1 or Figure 323.2.2.

Considering the examples of the preceding section, for 6 NPS Sch. 40 (0.280 in. WT) ASTM A 312 Type 316L seamless pipe, impact testing is not required at temperatures above -325°F, given that the ASTM standard ensures that carbon content is below 0.1% and that the material is in the solution treated condition. However, for welds joining the seamless pipe, the situation is different. Although, most austenitic stainless steel base metals are considered suitable for very low design minimum temperatures (e.g., -325°F), impact tests must be conducted on weld metal deposits for design minimum temperatures below -20°F, except in unusual cases where the notes to B31.3 Table 323.2.2 are applicable. One reason for weld metal impact testing is that the microstructure of most austenitic stainless steel weld metals is not fully austenitic. The chemical composition of most weld metal is purposely adjusted to provide some delta ferrite, which reduces the risk of hot (solidification) cracking during welding operations. Ferrite is, however, notch sensitive, and excessive ferrite can make an austenitic stainless steel weld metal behave in a brittle manner. Consequently, impact testing at or below the DMT is a method of guarding against excessive delta ferrite and brittle fracture, when the DMT is below -20°F.

With the previous 6 NPS Sch. 160 (0.719 in. WT), ASTM A 106 Grade B pipe example, if DMT is above the +15°F minimum permissible temperature for use of the metal without impact testing, then impact testing is not required. It is also not required for the weld metal. However, if the DMT were below +15°F, impact testing of the base metal would be required, provided that the 25%/6 ksi note of Table 323.2.2 is not applicable. In this case, it would probably be more effective to purchase pipe impact tested according to a pipe procurement standard (e.g., ASTM A 333 Grade 6 pipe), and if the DMT is below -20°F, weld impact testing would also be required. Weld metal and heat affected zone impact testing is normally done as part of the weld procedure qualification and need not be repeated for production welds (see Note 2 of Table 323.2.2).

Avoiding Low Temperature Materials

In some situations, it may be possible to avoid low temperature materials through an increase in the design minimum temperature. Justification for the increase could be based upon:

a) changes to the process;
b) additional heat available from the process;
c) use of insulation, with or without heat tracing; and/or
d) planned warm start-ups.

Another opportunity for avoiding low temperature materials is contained in Note 2 of Table 323.2.2. The note states: "Impact testing is not required if the design temperature is below -20°F but at or above -50°F, and the maximum operating pressure of the manufactured components will not exceed 25% of the maximum allowable design pressure at ambient temperature, and the combined longitudinal stress due to pressure, dead weight, and displacement strain (see ¶319.2.1) does not exceed 41 MPa (6 ksi)."

While it is sometimes possible to avoid low temperature materials, note that it is not always practical, cost effective, or technically appropriate. For example, additional costs for temperature permissive controls may be necessary to prevent pressurization below the design minimum temperature or to force shut-down in the event of a tracing failure. Engineering costs necessary to develop arguments against the need for low temperature materials can quickly erode any potential savings created by avoiding impact tested materials. Finally, avoidance of brittle fracture in any specific design situation may involve a lot more than simple consideration of DMT, impact test temperature, and impact test acceptance criteria.

Common Code Paragraphs Relating to Notch Toughness and Low Temperature Requirements

For the convenience of users of this guide, several clauses and tables applicable to B31.3 impact testing requirements are listed in Table 5.15 below. Users are cautioned that this guide is not a substitute for the ASME B31.3 Code which should be consulted for all requirements affecting pressure piping design and construction.

Table 5.15 B31.3 Clauses and Tables Applicable to Impact Testing

Clause or Table	Description
301.3.1	Design Minimum Temperature
301.5.1	Dynamic Effects - Impact
301.9	Reduced Ductility Effects
302.2.4(h)	Allowances for Pressure and Temperature Variations (below the minimum temperature shown in Appendix A)
309.2.2	Carbon Steel Bolting (note that B7M, L7M, B7, L7, 2HM, 7M, 2H, and 7 are low alloy steels, not carbon steels)
321.1.4(c)	Materials (of unknown specification used for piping supports)
323.2	Temperature Limitations
323.2.2	Lower Temperature Limits, Listed Materials
323.2.3	Temperature Limits, Unlisted Materials
323.3	Impact Testing Methods and Acceptance Criteria
323.4.2(a)	Ductile Iron
323.4.2(b)	Other Cast Irons
Table 323.3.1	Impact Testing Requirements for Metals
Table 323.2.2	Requirements for Low Temperature Toughness Tests for Metals
Table 323.3.5	Minimum Required Charpy V-Notch Impact Values
Table A-1	Basic Allowable Stresses in Tension for Metals (see Min. Temp. column and Note 6 at the beginning of the table)
Table A-2	Design Stress Values for Bolting Materials (see Min. Temp. column and Note 6 at the beginning of the table)

Materials Selection

When selecting materials for plant piping systems, legal, code, commercial, and technical considerations must be addressed.

Legal Considerations

Legal considerations include understanding and appreciation of:

a) legislation applicable to the jurisdiction having authority over the design, construction and operation of the piping system; and
b) contracts which exist between various parties.

For example, many states, provinces, and countries have legislated the use of B31.3 rules for construction of piping systems, so the Code essentially becomes a legal document. Local jurisdictions may also operate under government acts and regulations which impose additional requirements that may be:

a) technical (e.g., prohibitions concerning certain materials or design practices), or
b) organizational (e.g., registration of designs, registration of welding procedures, registration and accreditation of quality control systems)

For many projects, contract documents, including specifications prepared by the owner or owner's engineer, impose restrictions on:

a) materials (e.g., quality level, notch toughness properties, service environment, product form), and
b) fabrication/construction methods (e.g., bending, forming, welding, heat treating, hydrotest water quality, cleaning chemicals)

B31.3 Code Considerations

As indicated earlier in this chapter, the B31.3 Code is concerned with pressure integrity (safety). This is manifest, for example, through provision of allowable design stresses as a function of temperature, rules for notch toughness evaluation and brittle fracture avoidance, restrictions for various fluid service categories, requirements for weld procedure qualification, restrictions on forming and bending practices, examination requirements, and numerous prohibitions, limitations, conditions, and precautionary measures scattered throughout the Code.

Although Code issues must be considered in the material selection process, the Code does not instruct the user on how to select specific materials. ¶300(c)(6) states: "Compatibility of materials with the service and hazards from instability of contained fluids are not within the scope of this Code. See para. F323." The first sentence of ¶F323(a) states: "Selection of materials to resist deterioration in service is not within the scope of this Code." Clearly, the technical issues related to materials selection must be considered by personnel with specific training in this area.

Commercial Considerations

Materials decisions invariably have an impact on project cost and schedule. Can the material be purchased in the required form? What will be the initial cost? When will the material be available? What is the life cycle cost relative to other material options? Who can fabricate the material into a piping system? When can it be delivered and at what cost? What are the anticipated maintenance costs with the selected material?

These questions can be difficult to answer, but every materials decision has a commercial impact which must be considered. The level of detail given to study of the commercial impact is widely variable depending on project scope, technical complexity, and management.

Technical Considerations

Technical considerations begin with an understanding of process and containment requirements including pressure, temperature, fluid velocity, and fluid characteristics, as well as risk and consequences of failure. Material properties are then evaluated in light of this understanding, and generally include consideration of chemical, mechanical, and physical properties, as well as corrosion resistance, weldability, and formability. Fundamentally, final materials selection involves the proper matching of material properties with process design conditions, mechanical design conditions, fabrication and construction conditions, and operating conditions. Table 5.16 provides an overview of some material selection issues relative to other project variables.

Table 5.16 Technical Materials Selection Dependence on Project Variables

Project Variable	Materials Selection Issues
Process Design Conditions and Corrosion Resistance	• general weight loss corrosion • localized corrosion (pitting, crevice, fretting) • stress corrosion cracking (a.k.a. environmental cracking)
Mechanical Design Conditions and Mechanical Properties	• strength (i.e., required thickness) • ductility and resistance to brittle fracture • hardness (requirements usually linked to process design via stress corrosion cracking issues) • low temperature notch toughness (in addition to ambient temperature effects, can also be linked to process design such as significant Joule-Thompson effects on blowdown of high pressure gases, or handling of cryogenic fluids) • high temperature strength and creep resistance • vibration and fatigue resistance
Fabrication and Construction Conditions	• shop or field work, including availability of equipment and skilled personnel • forming and bending needs • joining needs (welding, threading, brazing, and soldering)
Operating Conditions	• ease of maintenance and repair, including availability of equipment and skilled personnel (e.g., manned or unmanned plant) • product delivery requirements, planned on-stream efficiency, and proposed plant or unit turnaround schedule • adequacy of project safeguards - e.g., prevention of leakage between streams (one or two check valves), adequacy of pressure and temperature controls

From Table 5.16, one might correctly assume that selection of materials is an evolutionary process, especially for technically complex projects. Generally the materials selection process involves a series of steps, separated in time, with each step narrowing down possibilities until the best option is determined. As project details unfold, this narrowing of options is typically manifested by a change in the way materials are identified, going from very generic descriptions (e.g., stainless steel) to specification, type, and grade (e.g., ASTM A 312 TP304L).

The level of detail applied during materials selection depends heavily on factors such as the scope, size, and complexity of the project, the stage of the project, the corporate style of the participants, and the availability of materials specialists. Although the steps sometimes overlap, careful analysis of several projects would show at least three distinct phases of materials selection with the following characteristics:

a) Conceptual design - overview of process technology
b) Process design - details of process technology
c) Mechanical design - primary scope of B31.3

Conceptual Design - Overview of Process Technology

Development of a materials selection and corrosion control philosophy for a project normally begins with evaluation and understanding of the process technology. Although this step in the materials selection process is not always evident or obvious, and may even be omitted when the process is well understood, it is a logical starting point for a grass roots project.

As an example, in the very early stages of project development, where "ballpark" economics are being thrashed out, the main interest of management is "big picture" material costs. That dreaded classic question generally reaches the materials specialist. How much stainless steel do we have to buy?. Of course, at the conceptual stage process details can be sketchy at best, so there is much crystal ball gazing at this point. Nevertheless, best efforts are made to provide a very generic overview of material options, knowing that management doesn't generally understand that there are at least fifty different kinds of stainless steel, each with its own unique price and delivery terms, and hundreds of material options that go by much stranger sounding names than "stainless steel."

The job of the materials specialist, who may also be called the corrosion specialist, is to examine the proposed process for any major issues which could jeopardize the project in financial and/or schedule terms. This is normally done in consultation with process professionals who have defined the process in terms of block flow diagrams, process stream compositions, and chemical reactions that may occur. A few typical questions to be answered during conceptual design are listed below.

a) Are carbon steels adequate?
b) Can coatings or chemicals be used economically to control corrosion?
c) If high alloy materials are required, how much will they cost?
d) How well is the interaction between the process environment and the material understood?
e) Is there a requirement for materials testing prior to selection?
f) Could required materials testing jeopardize the project in financial or schedule terms?

As an example, consider conceptual design for a plant for handling high velocity wet natural gas at 1000 psi, with 25% CO_2 and 2% H_2S at 200°F. Bare carbon steel would not be a likely candidate material due to excessive weight loss corrosion. Inhibitors could be considered, but adequate corrosion control may not be possible. The process is a bit hot for conventional organic coatings or plastic liners. Both CO_2 and H_2S could permeate the coating and cause coating failure if the system were subject to rapid pressure changes. Intermediate alloys containing chromium could work with adequate control welding and heat treatment. High alloy steels (stainless steels) could also work, but there is a large increment in cost and there may be potential for chlorides and resultant chloride stress corrosion cracking (ClSCC).

Process Design

If development economics favor further work, the next phase of material selection is generally made coincident with or slightly lagging behind process design. Process information is examined in detail and materials are selected based on compatibility with process stream characteristics and other external variables if they are known. Decisions often require consultation with process engineers for clear understanding of process conditions, including steady state and non-steady state conditions such as start-up, upset, planned shutdown, and unplanned shutdown.

As an example of the material selection process for a given stream, one might designate piping materials such as carbon steel (CS) with a suitable corrosion allowance (CA), if no characteristics other than low cost and moderate resistance to corrosion were important. However, if it is also known that the pipe might be exposed to low ambient temperatures where impact properties were required, one might upgrade the material selection to CSIT (carbon steel, impact tested). For another stream, which could be considered very corrosive, one might specify 3XXL stainless steel, where "L" grade is imposed to resist HAZ sensitization during welding, and the possibility of intergranular corrosion in service. If that same stream contained an aqueous chloride phase at 80°C, 3XX would be susceptible to ClSCC. In that case, one might specify a duplex stainless steel or superduplex stainless steel, depending upon chloride level, oxygen content and temperature.

While there is no single method for making and documenting materials selection decisions, the normal output at this stage of a project is a Corrosion and Materials Report. The word *corrosion* is generally included in the title of the report, since many of the material selection decisions reflect a response to corrosion predictions. The report typically contains the basis for decision making, as well as narratives describing issues, concerns, and limitations governing final materials selection for a given portion of the process.

Material selection diagrams (MSD's) and/or material selection tables are generally included with the Corrosion and Materials Report. MSD's are typically modified process flow diagrams (PFD's) showing generic material choices and corrosion allowances for each corrosion circuit. Corrosion circuits are elements of the process with similar corrosion characteristics, and are frequently equivalent to process streams defined on the PFD, or to subsets of process streams. Presentation of materials selection data in diagrams is generally the most useful format for communication with other design professionals, who will use the information during completion of subsequent work. Such presentation also assist in maintaining materials engineering input on the project. A simplified example of a MSD for an amine sweetening unit is shown in Figure 5.1.

C-201
Amine Contactor
Shell: CS + 1/8" CA + HIC
Heads: CS + 1/8" CA + HIC
Nozzles: Seamless CS + 1/8" CA

F-201
Lean Amine Filter
Shell: CS + 1/8" CA

V-201
Rich Amine Flash Tank
Shell: CS + 1/8" CA
Heads: CS + 1/8" CA
Nozzles: CS + 1/8" CA

C-202
Amine Regenerator
Shell: CS + 1/8" CA
Shell: CS + 1/8" 3XXL
Heads: CS + 1/8" 3XXL
Nozzles: To Match Shell/Head
(Cladding As Shown On Sketch)

V-202
Reflux Accumulator
Shell: CS + 1/16" + IPC
Heads: CS + 1/16" + IPC

P-201 A/B
Lean Amine Pumps
Case: CS
Impeller: CI

E-201
Lean Amine Cooler
Headers: CS + 1/8" CA
Tubes: CS

E-202
Rich/Lean Exchanger
Shell: CS + 1/8" CA
Head: CS + 1/8" CA
Tubes: 3XXL
Tubesheet: CS
Channel: CS+1/8" CA
Channel Cover: CS + 1/8" CA

E-203
Stripper Overheads Condenser
Headers: 3XXL + 1/32" CA
Tubes: 3XX

P-202 A/B
Reflux Pumps
Case: CS
Impeller: CI

E-204
Reboiler
Shell: CS + 1/8" CA
Heads: CS + 1/8" CA
Tubes: CS
Tubesheet: CS + 1/8" CA

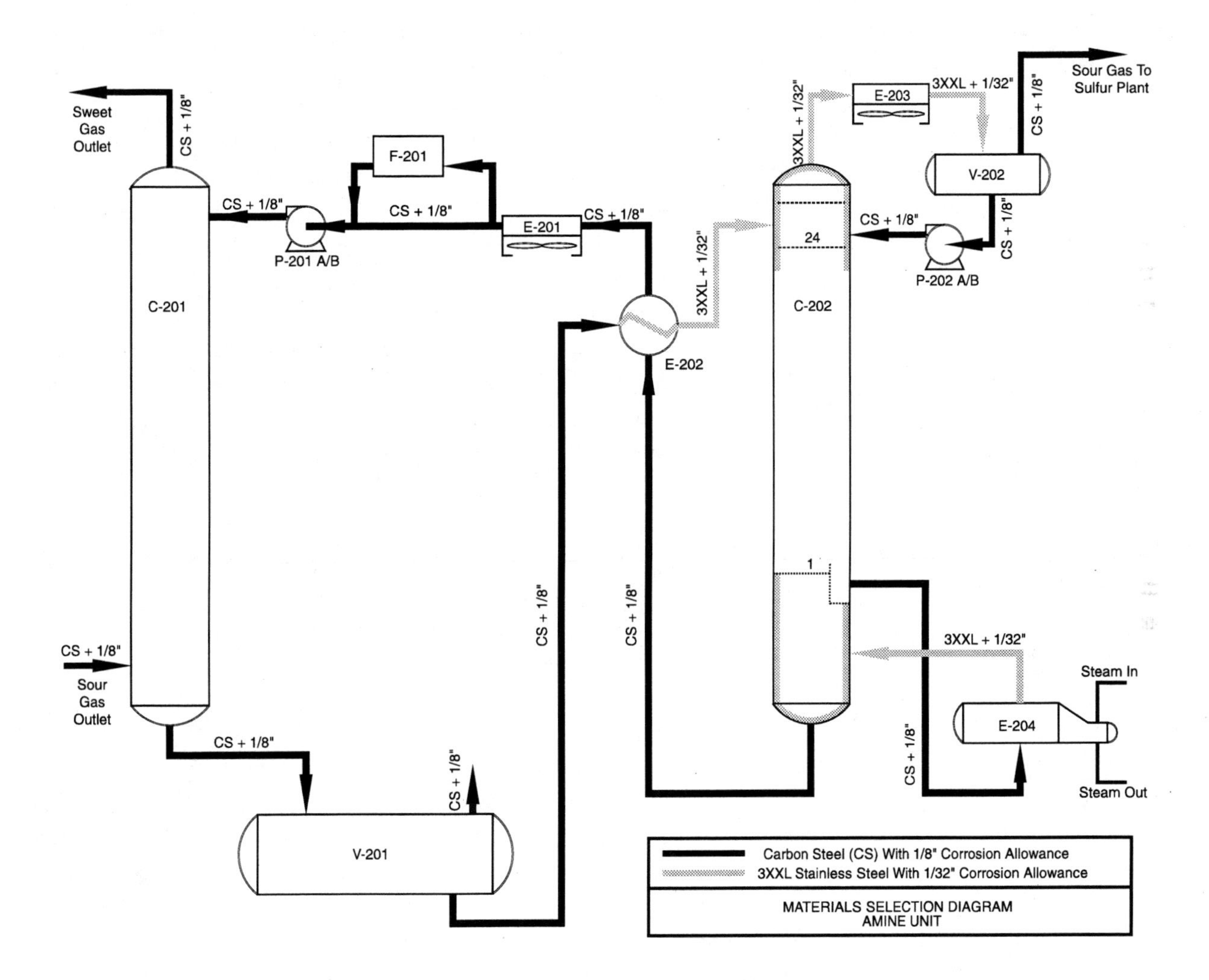

Note: This simplified Material Selection Diagram (MSD) is included in this book for illustration purposes only. Material selection for any process unit must be made on a project specific basis. MSD's usually include or reference the process information upon which material decisions are based (e.g., Pressure, Temperature, Velocity, and Stream Composition).

Figure 5.1 Simplified materials selection diagram (MSD).

Mechanical Design

During the mechanical design stage, the process characteristics shown by process simulations and PFD's, and the material requirements shown by the Corrosion and Materials Report and MSD's, are carefully examined. Then a *piping class*, which summarizes detailed mechanical and material requirements, is generally assigned to each line in the piping system. Note that the use of piping classes is not a B31.3 requirement; it is a typical approach to mechanical design within the pressure piping industry.

An example of a simple piping class is shown in Figure 5.2. It is a table of information for a given pressure rating and service characteristic, which describes acceptable piping component types and dimensions, as well as material specifications, types, classes, and/or grades. Piping components include pipe, forgings, fittings, valves, bolting, gaskets, and other piping specials. Usually the piping class contains information about corrosion allowance, postweld heat treatment, and nondestructive examination. Within a corporation, piping classes often exist from previous projects, or are supplied by the firm contracted to complete the engineering. When they do not exist, they are usually developed on the basis of process, material, and mechanical requirements.

The piping class designator is typically an alphanumeric descriptor such as D2 or AA2U, which is shown on the piping drawings (collectively, a group of drawings including P&ID's, MFD's, piping plans and sections, and isometrics), usually as part of the line number. A typical line number would be 6 in.-HC-34212-D2, where 6 in. is the nominal pipe size, HC is the commodity descriptor (e.g., hydrocarbon), 34212 is the line serial number, and D2 is the piping class. Specification breaks, which may be for material or pressure reasons, are also applied to piping drawings.

Depending on project scope and technical complexity, the mechanical design stage may also include preparation of detailed material specifications (stand alone or supplementary) to address issues such as material chemistry, processing requirements, product form (cast, forged, welded), fracture toughness, weldability, heat treatment, nondestructive examination, and various forms of corrosion resistance.

For some projects it has been said, sometimes in rather unkind words, that there is no need for materials engineering input. Piping designers can simply work from available prepackaged piping classes. Although there may be some merit in minimizing materials engineering input, remember that someone had to develop the prepackaged piping classes, and it was probably someone with extensive materials knowledge. Consequently, it may be useful to involve a materials specialist, even if it is just to bless the mechanical designer's cookbook (e.g., piping classes).

Service:	Sweet Hydrocarbons			**Piping Class:**	A1
Material:	Carbon Steel		**ASME/ANSI B16.5 Pressure Class:**		150
Radiography:	10% of Butt Welds		**Temperature Limits (1):**	-29 to 200°C (-20 to 400°F)	
PWHT:	Per B31.3		**Corrosion Allowance:**	1.6 mm (0.0625 in.)	
Item	**Size (NPS)**	**Thickness Or Rating**	**General Description**	**Material Spec And Grade**	**Notes**
Pipe					
	¾ - 1 ½	Sch. 80	PE, Seamless	A 106 Gr. B	
	2 - 18	Sch. 40	BE, Seamless	A 106 Gr. B	
BW Fittings					
45° Elbow:	2 - 18	Sch. 40	BE, BW, Seamless	A 234 Gr. WPB	
90° Elbow:	2 - 18	Sch. 40	BE, BW, Seamless	A 234 Gr. WPB	
Reducing Tee:	2 - 18	Sch. 40	BE, BW, Seamless	A 234 Gr. WPB	
Straight Tee:	2 - 18	Sch. 40	BE, BW, Seamless	A 234 Gr. WPB	
Conc. Reducer:	2 - 18	Sch. 40	BE, BW, Seamless	A 234 Gr. WPB	
Ecc. Reducer:	2 - 18	Sch. 40	BE, BW, Seamless	A 234 Gr. WPB	
Forged Fittings					
Sockolet:	¾ - 1½	3000	SW, FS	A 105	
Threadolet:	¾ - 1½	3000	Thr'd, FS	A 105	
Weldolet:	2 - 6	Sch. 40	BW, FS	A 105	
Union:	¾ - 1½	3000	SW or Thr'd, FS	A 105	
Flanges					
Weld Neck:	¾ - 1½	300	RF, Forged Steel	A 105	
	2 - 18	300	RF, Forged Steel	A 105	
Blind:	¾ - 18		RF, Forged Steel	A 105	
Fasteners					
Studs:	All		Low alloy steel	A 194 Gr. B7	
Nuts:	All		Low alloy steel	A 194 Gr. 2	
Gaskets:	2-18		Spiral Wound	API 601 304 SS	
Valves:					(2)
Gate:	1/2 - 1½	800	SW	VG-801	
	1/2 - 1½	800	SW*Thr'd	VG-803	
	2-18	150	RF	VG-101	
Ball:	1/2 - 1½	2000 CWP	Thr'd, Split Body	VB-2002	
	2-10	150	RF	VB-131	
Check	¾ - 1½	800	Thr'd	VC-802	
	2 - 4	300	RF, Swing Type	VC-101	
	6 - 18	300	RF, Wafer Type	VC-121	

(1) For thickness above 0.500 in. (12.7 mm) see B31.3 Fig. 323.3.2 regarding impact test requirements.
(2) See owners Piping Class Index for key to valve designations VG-801, etc.

Note: PE = Plain End, BE = Bevel End, SW = Socket Welding, BW = Butt Welding, Thr'd = Threaded, FS = Forged Steel, CWP = Cold Working Pressure

Figure 5.2 Layout and Content of a Typical Simple Piping Class.

Material Certificates

As a project progresses from design to material procurement and construction, it is generally desirable to obtain confirmation that incoming materials satisfy engineering requirements. This is usually accomplished through a review of documents which describe material characteristics, generally in accordance with standards or specifications stated on the purchase order. Typically these documents are known as Mill Certificates, Test Certificates, Mill Test Certificates, Certificates of Compliance, and Material Test Reports (MTR's).

A sample Test Certificate is shown in Figure 5.3. It describes the chemical and mechanical properties obtained during tests conducted to satisfy the requirements of ASTM A-105. During review of the document, data shown is compared to ASTM A-105 to verify compliance with requirements. Quality control checks during receiving of the flange should be used to verify flange dimensions, rating and heat identification, which should be stamped on the flange.

HOTHEAD FORGINGS LTD

1234 Just North Avenue
Somewhereville, Alberta

Customer:	Alberta Rig Welder Supplies Inc.
PO No.:	ARWS 0101234
Description:	ASME B16.5 Flanges, 2 NPS, 600#, RFWN, Sch. XXS
Material:	A-105

	Heat		**Chemical Analysis**									
Item	**Code**	**No.**	**C**	**Mn**	**P**	**S**	**Si**	**Ni**	**Cr**	**Mo**	**Cu**	**V**
1	X-857	12-084572	0.177	1.12	0.012	0.014	0.27	0.02	0.01	0.003	Tr.	<0.01
2												
3												
4												
5												
6												
7												
8												
9												
10												

	Mechanical Properties				**Charpy Impact Test**					
Item	**Tensile**	**Yield**	**Elong.**	**R of A**	**Size**	**Temp**	**S1**	**S2**	**S3**	**Avg.**
1	73220	42519	27.6	55.4						
2										
3										
4										
5										
6										
7										
8										
9										
10										

Conditions
Normalized at 900°C for 1 hour. Cooled to room temperature in still air. Hardness = 161 HB.

I hereby certify that the contents of this document are correct according to the records in possession of this company.

QA Department: ______________________________

Date: ____________________

Figure 5.3 Typical Mill Test Certificate.

Chapter 6

FABRICATION, ASSEMBLY, AND ERECTION

Introduction

Chapter V of the B31.3 Code is devoted to the fabrication, assembly, and erection of piping systems. These terms are defined by ¶300.2 as follows.

a) *Fabrication* is the preparation of piping for assembly, including cutting, threading, grooving, forming, bending, and joining of components into subassemblies. Fabrication may be performed in the shop or in the field.
b) *Assembly* is the joining together of two or more piping components by bolting, welding, bonding, screwing, brazing, soldering, cementing, or use of packing devices as specified by the engineering design.
c) *Erection* is the complete installation of a piping system in the locations and on the supports designated by the engineering design, including any field assembly, fabrication, examination, inspection, and testing of the system as required by the Code.

Fabrication, assembly, and erection require the use of many special processes including:

a) forming and bending by cold and hot methods;
b) joining by welding, brazing, soldering, or mechanical methods including threading, flanging, specialty high pressure connections, and mechanical interference fits (MIF); and
c) heat treatment by local methods, or by permanent or temporary furnaces.

B31.3 assumes some understanding of the special processes used during fabrication, assembly, and erection of piping systems. However, as with materials of construction, the level of understanding is widely varied and often restricted to a few processes in the user's repertoire of experience. Consequently, the objective of this chapter is to explore the basic technology behind some of the special processes in relation to requirements of the Code.

Bending and Forming [¶332]

General

Bending and forming processes are sophisticated technical operations. An evaluation of the effects of bending and forming on material properties is integral to the use of such products in piping systems. In this light, the following Code statements should be considered as more than simple motherhood:

a) ¶332.1 states: "Pipe may be bent and components may be formed by any hot or cold method which is suitable for the material, the fluid service, and the severity of the bending or forming process."
b) ¶332.3 states: "The temperature range for forming shall be consistent with material, intended service, and specified heat treatment."

These Code clauses are intended to trigger the engineering input necessary to verify that final material properties will be satisfactory for the intended service. And, even though the Code does impose requirements for design (e.g., ¶304.2) and fluid service (e.g., ¶306.2), engineering input is still needed. The Code does not and can not provide rules to address the specific requirements of every situation.

As part of an engineering evaluation, below are some useful starting questions regarding the effects of bending and forming on material properties for a specific service.

a) What effect will the bending or forming temperature and deformation parameters (e.g., cold, warm, or hot bending, strain rate and total strain) have on strength, ductility, hardness and notch toughness of the resulting bend?
b) What effect will the resulting microstructure have on general corrosion, localized corrosion (galvanic, pitting, and/or crevice corrosion), stress corrosion cracking, or long term mechanical properties?
c) What are the risks relative to formation of hard spots, undesirable precipitation effects, fatigue resistance, and creep resistance?

Hopefully the above questions would be answered with the help of metallurgical and/or corrosion specialists, in combination with suitable testing when appropriate.

Bending

The need for changes to the direction of flow in piping systems has traditionally been accommodated through the use of manufactured fittings such as elbows and tees. However, changes to direction of flow may also be made through the use of pipe bends. In fact, with modern equipment, substantial economic benefits can be derived from the use of bends, by virtue of reduced fitting, welding, and nondestructive examination (NDE) costs.

Before examining bends in detail, a few comments regarding bend types may be useful to readers with no bending experience. In the bending and piping industries, bend types are often described by a multitude of terms. Although a formal classification system does not exist, bends are usually referred to in terms of:

a) method of manufacture, including cold bends, hot bends, furnace bends, induction bends, arm bends, ram bends, three point bends, miter bends, segmented bends, corrugated bends, and creased bends;
b) location of manufacture, that is, field bends or factory bends;
c) shape or appearance, such as L-bends, S-bends, wrinkle bends, miter bends, segmented bends, corrugated bends, and creased bends; and
d) function or end use, such as sag bends, overbends, side bends, and combination bends.

Note that several of the terms described in (c) and (d) above are rooted in cross-country pipeline construction, where bends are normally used to accommodate changes of elevation associated with the terrain or to provide for expansion and contraction of the pipeline with changes of temperature. For plant piping systems, it is most common to use bending terms reflective of the method of manufacture which may include combinations of terms (e.g., three point cold bend, hot furnace bend, hot induction bend).

Regardless of bend type, all bends have certain features and dimensional characteristics which must be carefully specified during piping design and controlled during bend procurement and manufacture (see Figure 6.1).

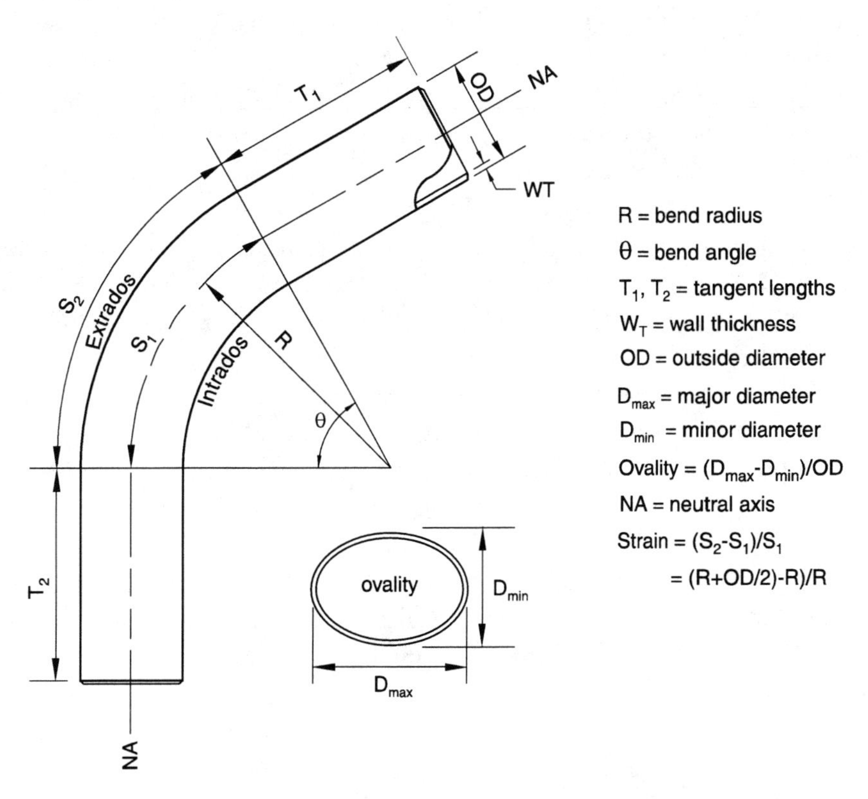

R = bend radius
θ = bend angle
T_1, T_2 = tangent lengths
W_T = wall thickness
OD = outside diameter
D_{max} = major diameter
D_{min} = minor diameter
Ovality = $(D_{max}-D_{min})/OD$
NA = neutral axis
Strain = $(S_2-S_1)/S_1$
= $(R+OD/2)-R)/R$

Figure 6.1 Typical pipe bend features and dimensional characteristics.

Listed below are some issues to consider when procuring bends:

a) Bend radius must be stated for each pipe size. This is frequently done in terms of the diameter of the pipe used to make the bend (e.g., 10D), but it may also be stated in specific numerical terms (e.g., R=100 in.). When stating bend radius as a function of pipe diameter, the diameter of reference should also be stated (e.g., nominal or outside). For example, a 10D bend in a piece of NPS 6 pipe would have a bend radius of 60 inches (i.e., 10 x 6 = 60 in.) based on nominal pipe diameter, but it would have a radius of 66.25 inches (i.e., 10 x 6.625 = 66.25 in.) based on outside diameter. And, while it maybe customary to specify bend radii in terms of nominal pipe diameter, all too often, the custom at one location is not the same as the custom at another location.
b) Tangent lengths and tolerances should be stated on the order. Tangent lengths should be computed to minimize the number of subsequent shop or field welds required in the piping system and should consider the maximum dimensions permitted for transportation of bends or assembled piping to the job site. Also, note that minimum tangent lengths may be required for bend manufacturing purposes (e.g., to grip the ends of the pipe or to push the pipe through a bender). This can affect the minimum length of pipe procured for bending purposes.
c) Ovality, or flattening as described by B31.3, is limited [¶332.2.1] to:
d) 8% for bends subject to internal pressure and
e) 3% for bends subject to external pressure.
f) Note that B31.3 does not permit removal of metal to satisfy ovality (flattening) tolerances. Also, tighter tolerances may be required for bends requiring passage of internal inspection tools.
g) ¶332.1 indicates that bends shall be substantially free from buckling (also referred to as wrinkling), but no tolerances are given. This topic should be an item of agreement between the purchaser and the bend manufacturer.
h) ¶332.2.2 defines cold and hot bending in terms of the transformation range for the material. With these definitions, the Code is assuming that users have some understanding of metallurgy. For carbon steel, the lower transformation temperature is about 1333ºF. The upper transformation temperature depends mainly upon carbon content and theoritically can be anywhere between about 1333ºF and 1675ºF. In more practical terms, cold bending of carbon and low alloy steels is carried out at temperatures in the black heat range where the material is ferritic in structure. Hot bending is carried out in the bright red heat range where the metal is austenitic in structure. For low alloy, intermediate alloy, and high alloy transformation hardenable steels, microalloyed steels, non-transformation hardenable steels, and non-ferrous metals and alloys, specialist help should be obtained when evaluating bending parameters (e.g., consider effects of strain hardening, precipitation, loss of corrosion resistance, and embrittlement). Care should also be exercised when bending steels in the blue brittle range (300 to 700ºF).
i) ¶332.1 states that thickness after bending or forming shall not be less than that required by the design. No maximum is imposed for thickening and this is common to plant piping situations. For some applications, however, engineering should consider the necessity of imposing a maximum thickness (e.g., to permit smooth passage of internal inspection tools and pigs commonly used in pipeline applications). In some jurisdictions, certain portions of pipeline systems have been constructed according to B31.3 rules.

j) B31.3 addresses limitations on outer fiber elongation (strain) in clauses dealing with post bend heat treatment [¶332.4]. In addition to B31.3 requirements, be cautious about outer fiber strain restrictions imposed by other standards which may be applicable to the piping system (e.g., NACE MR0175).
k) Although not specified in B31.3, when longitudinally welded pipe is used for bending, the longitudinal weld should be placed as near as practical to the neutral axis of the bend.

Heat Treatments Required After Bending or Forming

Through heat treatment rules, B31.3 does address some of the adverse effects of bending and forming operations on material properties. The rules are based on type of bending operation (hot or cold), type of material, and outer fiber strain.

Regardless of material thickness, heat treatment is required according to conditions prescribed by ¶331:

a) after hot bending and forming operations carried out on P Nos. 3, 4, 5, 6, and 10A materials;
b) after cold bending and forming:
   i) of P-No. 1 to P-No. 6 material where the outer fiber elongation in the direction of severest forming (usually extrados) exceeds 50% of specified minimum elongation stated for the specification, grade, and thickness of the starting pipe material;
   ii) of any material requiring impact testing, if the maximum calculated fiber strain after bending or forming exceeds 5%; and
   iii) when specified by the engineering design.

With regard to (a) above, the materials listed are capable of transformation hardening when cooled from hot bending and forming temperatures, so heat treatment is aimed at restoring mechanical properties to a level reasonably consistent with the starting materials. With regard to (b) above, work hardening effects imparted by cold bending and forming reduce the ductility and notch toughness (impact strength) of materials. As well, cold bending and forming operations generate residual stresses in the finished parts. Heat treatment is therefore applied as a tool to reduce the negative consequences of these effects (e.g., brittle fracture).

Bending References

Pipe Fabrication Institute (PFI) Standard ES-24 is an excellent starting reference for pipe bends and includes information on terminology, bending methods, tolerances, material allowances, and metallurgical precautions.

Welding

Most people associated with the pressure piping industry will, at some point, come in contact with welding. This could mean:
(a) writing a welding procedure,
(b) qualify a welding procedure or welding personnel,
(c) reviewing a welding procedure for acceptance or rejection in a specific application, or
(d) doing the welding, which will likely involve trying to interpret someone else's welding procedure.

To laymen, welding is a magic act. With the common arc welding processes, the arc ignites with a flash of bright light, the magician (welder) moves the bright light along the interface between the metals and "presto," a weld is created. Of course, well informed piping professionals know that welding is a lot more than smoke and mirrors. It is a very complex interdisciplinary science involving aspects of mechanical, civil, electrical, and metallurgical engineering. Thorough technical understanding of welding operations requires specific training, which is only available at select engineering, technical, and trade schools.

B31.3 provides welding guidance in the areas of:

a) responsibility [¶328.1],
b) qualifications [¶328.2], and
c) technical and workmanship criteria [¶328.3 through ¶328.6].

These are discussed in detail in the following sections.

Welding Responsibility [¶328.1]

B31.3 is very clear regarding responsibility for welding. ¶328.1 states: "each employer is responsible for the welding done by the personnel of his organization and, except as provided in paragraphs. 328.2.2 and 328.2.3, shall conduct the tests required to qualify welding procedures, and to qualify and as necessary requalify welders and welding operators." This philosophy is consistent with other sections of the ASME code and with similar codes, standards, and specifications around the world.

The two exceptions in ¶328.1, regarding the need for employers to conduct welding qualification tests, are:

(a) Procedure Qualifications By Others [¶328.2.2], and
(b) Performance Qualification By Others [¶328.2.3].

Although these exemptions exist, the employer is not exempt from responsibility for welds prepared according to procedures or personnel qualified by others. Close examination of the conditions attached to these exemptions will enhance understanding as to why the employer is responsible for all welding.

Regarding exemption of an employer from weld procedure qualifications through use of welding procedures qualified by others, interpretation of B31.3 restrictions shows that the exemption is limited to low risk situations. For the exemption to apply, the Code has the following requirements:

a) The inspector must be satisfied [¶328.2.2(a)]:
   i) with the capability of the organization providing the procedure [¶328.2.2(a)(1)]; and
   ii) that the employer intending to use the procedure has not made any changes to it.
b) Base metals are restricted to P-No. 1, P-No. 3, P-No. 4 Gr. 1 (13%Cr max.), or P-No. 8 [¶328.2.2(b)].
c) Impact tests are not required [¶328.2.2(b)].
d) Base metals to be joined are of same P-No., except P-No. 1, P-No. 3, and P-No. 4 Gr. 1 may be welded to each other as permitted by ASME Section IX [¶328.2.2(c)].
e) Base metal thickness does not exceed ¾ in. (19 mm) [¶328.2.2(d)].
f) PWHT is not required [¶328.2.2(d)].
g) Design pressure does not exceed ASME/ANSI B16.5 Class 300 at design temperature [¶328.2.2(e)].
h) Design temperature is in the range -20ºF through 750ºF (-29ºC through 399ºC), inclusive [¶328.2.2(e)].
i) Welding processes are restricted to SMAW or GTAW or combination thereof [¶328.2.2(f)].
j) Welding electrodes are restricted to those listed in ¶328.2.2(g).
k) Employer accepts responsibility for both the WPS and PQR by a signature [¶328.2.2(h)].
l) Employer has at least one welder/welding operator who, while in his employ, has passed a performance qualification test using the procedure and the P No. of material specified in the WPS. Qualification must have been with a bend test per ASME Section IX Para QW-302. Qualification by radiography alone is not acceptable [¶328.2.2(i)].

Always consult the latest revision of the Code for any changes to the exemption restrictions, which may occur from time to time.

Regarding exemption of an employer from performance qualification testing, a performance qualification made by another employer may be accepted, subject to the following restrictions [¶328.2.3]:

a) The inspector must specifically approve the exemption.
b) The qualification is limited to piping using the same or equivalent procedure with essential variables within the limits of ASME Section IX.
c) The employer must obtain a copy of the performance qualification test record from the previous employer, which contains information prescribed by ¶328.2.3. This requirement seriously restricts the portability of welding qualifications. For example, why would an employer release such records to a competitor, unless the competitor is part of the same corporate group? The restriction has been overcome in some North American locations through qualification tests administered by the jurisdiction having authority over the work (e.g., state or province). In other locations, especially developing countries where massive projects are undertaken, the portability restriction has been overcome by general acceptance of qualification cards issued by some large local industrial organization Typically, the qualifying organization issues each

welder a laminated ID card containing the welder's photograph and other qualification information. Of course, welder performance qualification cards are valuable. Laminated photographic ID is essential and must be carefully controlled. For example, without laminated photographic ID it would be quite possible to have one qualified welder showing up at different test centers, passing qualification tests for all of his friends. Alternatively, it would be possible for unqualified welders to be substituted on the work by simple forgery of qualification cards without photographic ID.

Welding Qualifications [¶328.2]

Although the qualification exemptions discussed above may be used, normally the employer is required to conduct welding qualifications. B31.3 controls the details of welding qualifications by referencing:

a) ASME Section IX [¶328.2.1(a)],
b) supplementary technical requirements [¶328.2.1(b) through 328.2.1(f)], and
c) requirements for qualification records [¶328.2.4].

By external reference to ASME Section IX, the B31.3 Code takes advantage of a general working document governing welding qualifications for the entire ASME Code. Topics in the following paragraphs, which address welding variables listed in ASME Section IX, are indicated by the "QW" prefix to clause numbers.

Welding Processes [¶QW-252 through ¶QW-265]

There are well over 100 welding processes recognized by AWS Specification A3.0. Of these, about fifteen are listed in ASME Section IX, but only five arc welding processes are commonly used in modern pipe fabrication. As explained in the following paragraphs, these five basic processes are:

a) Shielded Metal Arc Welding (SMAW)
b) Gas Tungsten Arc Welding (GTAW)
c) Gas Metal Arc Welding (GMAW)
d) Flux Cored Arc Welding (FCAW)
e) Submerged Arc Welding (SAW)

It is worth mentioning that there are variations on some of these processes, related to method of filler metal addition (e.g., hot wire, cold wire) and metal transfer (e.g., short circuit, globular, spray). These are covered under the main process headings below. Plasma arc welding (PAW), with many of the characteristics of GTAW, is gaining popularity, but is not treated as a separate topic in this book.

Shielded Metal Arc Welding

Shielded Metal Arc Welding (SMAW) is an arc welding process which produces coalescence of metals by heating them with an arc struck between a flux coated metal electrode and the work. Shielding is obtained from decomposition (burning) of the electrode flux coating, which forms a gas shield around the end of the electrode and a slag covering over the hot and molten metal. The electrode contributes filler metal to the weld pool through melting of the core wire and metallic ingredients found in the electrode coating. Metallic ingredients in the electrode coating also contribute alloy elements in varying amounts, depending on electrode classification. The SMAW process is also known as Manual Metal Arc Welding (MMA) and stick welding.

Gas Tungsten Arc Welding

Gas Tungsten Arc Welding (GTAW) is an arc welding process which produces coalescence of metals by heating them with an arc struck between a nonconsumable tungsten electrode and the work. Protection of hot and molten metal from the atmosphere is achieved with an externally supplied shielding gas or gas mixture which flows from the gas cup (lens) surrounding the tungsten electrode. Filler metal may or may not be added to the weld pool. When filler metal is not added, the weld is called an *autogenous weld*. When filler metal is added, it is normally done manually by feeding a short length of "wire" into the weld pool. Sometimes filler metal is added automatically. For example, in a process known as GTAW Hot Wire Method, the wire is warmed by resistance heating prior to reaching the weld pool, resulting in increased productivity. Around the world, the GTAW process is also known by names such as TIG (Tungsten Inert Gas) welding, Heliarc welding, and argon arc welding. With a power supply capable of providing pulsed current, the process is termed Gas Tungsten Arc Welding -Pulsed Arc (GTAW-P).

Gas Metal Arc Welding

Gas Metal Arc Welding (GMAW) is an arc welding process which produces coalescence of metals by heating them with an arc struck between a continuous fed consumable electrode and the work. Protection of hot and molten metal is accomplished using an externally supplied shielding gas or gas mixture which flows from the end of the welding torch. Solid filler metal is supplied on spools of varying size suited to the type of application. Around the world, this process is known by a number of other non-preferred terms including MIG (Metal Inert Gas) welding, MAG (Metal Active Gas) welding, CO_2 welding, short-arc welding, dip transfer welding, and microwire welding.

Depending on welding current, voltage, type of shielding gas, and power supply characteristics, various types of metal transfer may be obtained. Gas Metal Arc Welding -Short Circuit Arc (GMAW-S) describes a process in which metal transfer from the consumable electrode occurs in repeated short circuits. This accounts for use of the non-preferred terms short-arc welding and dip transfer welding. Other forms of metal transfer include globular transfer and spray transfer (common for high production). Finally, when using a power supply capable of providing pulsed current, the process is termed Gas Metal Arc Welding -Pulsed Arc (GMAW-P). Detailed explanations of the mechanics of metal transfer may be found in specialized welding texts such as the *AWS Welding Handbooks*, published by the American Welding Society.

Flux Cored Arc Welding

Flux Cored Arc Welding (FCAW) is an arc welding process which produces coalescence of metals by heating them with an arc struck between a continuous tubular, consumable electrode and the work. The consumable electrode supplies filler metal to the joint by melting the tube portion of the electrode and by melting metallic ingredients in the flux core of the electrode. Hot and molten metal is protected from the atmosphere with a shielding gas produced by burning the core material inside the electrode, and with a slag covering produced by melting the core ingredients. Additional shielding gas may also be provided from an external supply which is fed through a conduit to the welding torch, where it is released into a shroud around the contact tip.

In many respects, the equipment used for FCAW is the same as the equipment used for GMAW. Consequently, FCAW is classified under the heading GMAW in ASME Section IX.

Submerged Arc Welding

Submerged Arc Welding (SAW) is an arc welding process which produces coalescence of metals by heating them with an arc struck between a bare metal electrode and the work. The arc is buried under a blanket of granular, fusible flux, hence the term "submerged arc" welding. During welding, a portion of the granular flux is melted by the arc to form a protective slag cover on the hot and molten metal. The electrode (welding wire) contributes filler metal to the joint, and this is sometimes supplemented by metallic ingredients found in the flux (minerals or metal granules). The SAW process is sometimes referred to as the Unionmelt process. In many cases, SAW can be set up for high production rates, using multiple electrodes and multiple arcs on the same welding head.

Power Sources for Arc Welding Processes

A detailed discussion of power sources is well beyond the scope of this book. However, the reader should appreciate that welding power sources are themselves, complex engineering works. Their function is to provide a stable, uniform supply of current in the type (AC or DC) and wave form best suited for the welding process, welding consumables, and base metals involved.

From a shop perspective, there are fundamentally two types of power sources:

a) constant current power sources (also known as droopers), and
b) constant voltage power sources (also known as constant potential power sources).

Generally, constant current (CC) power sources are used with manual welding (e.g., SMAW or GTAW) where variations of arc length are accommodated by a power source volt-ampere characteristic which keeps the current relatively constant. Constant potential (CP) power sources are used with the semi-automatic and automatic welding processes such as GMAW, FCAW, and SAW. In this case, constant arc length (i.e., constant voltage) is maintained by the power source by an increase or decrease of the electrode (wire) feed rate. With modern electronics, state-of-the-art multi-process power sources can have CP and CC built in to the same machine.

Joints [¶QW-402]

The terminology used to describe weld joints can be confusing, with meaning often dependent on the industry sector where work is carried out, as well as on applicable codes and standards. Experience and intuition often form the basis for understanding various terms. For example, the term "butt weld" is commonly used in pressure piping applications to describe a butt joint with a groove weld, which is typically a full penetration groove weld.

To properly communicate welding requirements, it is important to make some basic distinction relative to:

a) joint types,
b) weld types, and
c) joint geometry and end preparation.

Following from these basics, several supplementary details should also be considered, including:

a) degree of joint penetration - full or partial,
b) weld profile - convex or concave,
c) alignment and fit-up details,
d) backing type, if used, including weld metal and backwelded joints, and
e) access for welding (i.e., single welded joints, double welded joints)

Joint Types

There are five basic types of joints, as described in Table 6-1. Note that the joint is the interface between the members. This distinction is important, since it governs interpretation of weld symbols and thickness of the joint (e.g., for preheat purposes).

Table 6.1 Five Basic Joint Types

Butt Joint	A joint between two members aligned approximately in the same plane.
Corner Joint	A joint between two members approximately at right angles (90°) to each other.
Tee Joint	A joint between two members approximately at right angles (90°) to each other, in the form of a T.
Lap Joint	A joint between two overlapping members.
Edge Joint	A joint between the edges of two or more parallel or nearly parallel members.

Types of Welds

There are three basic types of welds, as listed in Table 6.2.

Table 6.2 Basic Weld Types

Groove Weld	A weld made in the groove between two members to be joined.
Fillet Weld	A weld of approximately triangular cross section, joining two surfaces approximately at right angles to each other.
Plug or Slot Weld	A weld made through a circular or elongated hole in one member of a lap or T joint, joining that member to the surface of the other member exposed through the hole.

End Preparation And Joint Geometry

End preparation [¶328.4.2] refers to the shape and dimensions of the base metal when viewed as a cross section at the welding end, prior to fit-up and welding. *Joint geometry* refers to the shape and dimensions of a joint when viewed in cross section after fit-up and prior to welding.

For pressure piping, joint geometry refers mainly to the type of groove to be used, of which there are many. B31.3 provides some guidance concerning end preparations and resultant joint geometry [¶328.4.2(a)(2) and Fig. 328.4.2]. For manually welded piping butt joints, the single-vee groove is most common. For thick joints, however, compound vee grooves may be used to reduce welding costs, distortion, and residual stress.

In other situations, decisions regarding joint geometry involve consideration of several issues, including Code requirements, selected welding process, size and type of filler metal, welding position, access to the root of the joint, availability of suitable tools for joint preparation, and economy.

Joint Penetration

Joint penetration refers to the minimum depth that a groove weld extends from its face into the joint, exclusive of reinforcement (i.e., weld metal in excess of the quantity required to fill the joint). With pressure piping, it is normal to aim for full (complete) penetration welds where the weld metal completely fills the groove and is fused to the base metal throughout the total thickness. Limits on incomplete penetration are listed in B31.3 and discussed later in this book.

In some cases, the engineering design may require partial penetration joints. The extent of penetration may be described in terms of the effective throat, although specification of the maximum amount of incomplete penetration is more useful when nondestructive examination methods are applied to verify joint quality. As an example of a situation where partial penetration joints would be used, consider a butt welded cement mortar lined piping system. With this type of construction, a thin gasket of suitable diameter (e.g., compressed asbestos or equivalent) is spot glued to the cement surface of one member of a joint. The other member is then fit up tight against the gasket and the joint is clamped. The root bead is then deposited, taking care not to penetrate completely through the root face of the carrier pipe. If the arc penetrates the root face of the carrier pipe, the gasket may burn and loose its ability to provide a seal, or concentrated heat from the arc may cause the cement to crack and fall away from the pipe wall.

Backing

Backing is a material placed at the root of a weld joint to support the molten weld metal [¶QW-492]. Backing may be metallic (e.g., metal rings or other weld metal) or nonmetallic (e.g., tapes with attached ceramic or refractory material) [¶328.3.2]. Backing rings may be manufactured as a continuous band or a split band [¶328.3.2].

When backing rings are used, B31.3 indicates that their suitability shall be demonstrated by procedure qualification, except that a procedure qualified without use of a backing ring is also qualified for use with a backing ring in a single welded joint [¶328.2.1(e)]. Note, however, that the welding technique used with a backing ring must provide for full penetration and fusion into the corner created at the junction of the pipe and backing ring. Even though there is an exception to further procedure qualification for single welded joints where backing is added, the welder must know how to achieve full penetration and fusion into this corner notch.

In general, the position of B31.3 regarding backing rings is consistent with ASME Section IX, which treats backing as a nonessential variable for the common arc welding processes. However, under strict interpretation, ASME classification of backing as a nonessential variable means that a WPS could be amended without requalification whether the backing were added or deleted. Under B31.3 rules, a deletion of backing from a single welded joint would not be permitted without requalification.

Regarding backing ring materials, B31.3 [¶328.3.2] provides restrictions which may be summarized as follows:

a) indicates that ferrous backing rings must be of weldable quality, with sulfur content not exceeding 0.05%. It is interesting to note that the phrase "weldable quality" has limited meaning in that there is no comprehensive definition of the term "weldable." It is also interesting to note that, while the 0.05% sulfur limit is aimed at minimizing the risk of hot (solidification) cracking, it is not very restrictive and would be of little value in controlling hot cracking susceptibility in high quality weld joints. The Code indicates that this sulfur restriction is applicable to ferrous metals, but it would be more appropriate to apply this to carbon and low alloy steels. For stainless steels, which are also ferrous materials, such restriction would be of negligible benefit in the prevention of hot cracking, especially in fully austenitic weld deposits.
b) addresses a requirement for weld procedure qualification of joints composed of ferritic and austenitic materials, which is in addition to ¶328.2.1(e). It is worth noting that the term ferritic includes a wide range of ferrous metals including carbon steels, low alloy steels, intermediate alloy steels, and the straight chromium stainless steels (ferritic and martensitic). Similarly,

while one normally thinks of 200 and 300 series stainless steels in the classification of austenitic materials, the group also includes many nonferrous alloys.

c) imposes two special requirements for nonferrous and nonmetallic backing rings, specifically:
   i) that the designer approve their use, and
   ii) that the welding procedure using them be qualified as required by ¶328.2.1(e).

In addition to the above Code restrictions on backing rings, designers normally prohibit their use for:

a) corrosive services where the space between the ring and the pipe might be a location for crevice or pitting attack,
b) cyclic or vibrating services where notches associated with rings become sites for development of fatigue cracks, and
c) cryogenic or low temperature service where notches become sites for initiation of brittle fracture.

Consumable Inserts [¶328.2.1(e), ¶328.3.3]

¶328.3.3 indicates that consumable inserts may be used provided:

a) they are of the same nominal composition as the filler metal,
b) they will not cause detrimental alloying of the weld metal, and
c) their suitability is demonstrated by weld procedure qualification [¶328.2.1(e)].

In general, the B31.3 position on consumable inserts is consistent with that of ASME Section IX, which treats consumable inserts as a nonessential variable for the common arc welding processes which would use inserts (e.g., GTAW). However, under strict interpretation, the ASME position on nonessential variables means that a WPS could be amended without requalification, whether the insert were added or deleted. Under B31.3 rules, addition or deletion of an insert would not be permitted without a supporting PQR (i.e., requalification).

Base Metals [¶QW-403]

Base metals and base metal designation systems have been discussed earlier in this book, and it should be clear that there are thousands of different base metal alloy systems. From a welding qualification perspective, it would be impractical to conduct tests each time a change of base metal occurs, so ASME developed a base metal classification system aimed at reducing the number of welding procedures required for a project. The system classifies base metals according to composition, weldability, and mechanical properties [¶328.2.1(f)]. The resulting system consists of:

(a) a set of P-Numbers (see Table 6.3), and
(b) a subset of Group Numbers for ferrous base metals, which controls situations requiring impact testing.

The underlying concept with respect to base metals and welding qualifications is simple. If you qualify a welding procedure by welding together two metals from the same P-Number, you qualify to weld any other metals assigned to that same P-Number classification (see ¶QW-424 for further explanation). For example, if you joined two pieces of ASTM A 106 Gr. B pipe, which are P-No. 1 materials, you are qualified to weld any other materials within the P-No. 1 material classification (subject, of course, to all of the other essential and supplementary essential variables of the code).

For ferrous metals which require impact testing, ASME Section IX imposes an additional restriction (i.e., a supplementary essential variable) on the range of base metals qualified, through a system of Group Numbers. If impact testing is required, the interchange of base metals is controlled by P-Number, and Group Number within the P-Number.

While the underlying concept of P-Numbers is simple, ASME Section IX clearly indicates that P-Number and Group Number assignments do not imply that any base metal may be indiscriminately substituted for a base metal used in the qualification test without consideration of compatibility from the standpoint of metallurgical properties, postweld heat treatment, design, mechanical properties, and service requirements.

Filler Metals [¶QW-404, ¶328.3.1, ¶328.3.3]

B31.3 states that filler metal shall conform to the requirements of ASME Section IX, except that a filler metal not yet incorporated in ASME Section IX may be used with the owner's approval if a procedure qualification test is first successfully made [¶328.3.1]. It should be noted, however, that ASME Section IX does not provide a list of "incorporated" filler metals. Instead, it provides a broad classification system for filler metals, set up on the basis of usability characteristics (by F-Number and AWS Classification) and weld metal analysis (A-Number). It is up to the Code user to determine whether proposed filler metals satisfy ASME Section IX. This is normally done by referring to literature of the filler metal manufacturer or to a chemical analysis of the weld deposit.

AWS Specifications and Classifications for Welding Consumables

There are literally thousands of welding consumables made by various manufacturers around the world. The American Welding Society (AWS) has devised a scheme for classifying welding consumables, which currently consists of some 30 different specifications. Information on welding consumables is also available in *The Metals Blue Book*™ - Welding Filler Metals, published *CASTI* Publishing Inc.

Table 6.4 lists all of the AWS specifications applicable to classification of welding consumables, which collectively includes consumable and nonconsumable welding electrodes, welding rods, and other filler metals. Appendix 1 outlines the alphanumeric classification system used in each AWS specification.

Table 6.3 List of ASME P-Numbers and Their Generic Descriptors

P-No.	Generic Descriptor
P-1	Carbon Steel, Carbon-Manganese Steel, Carbon-Manganese-Silicon Steel
P-2	Wrought Iron (No Longer Used)
P-3	Low Alloy Steel (12% total alloy typically)
P-4	Low Alloy Steel (Cr-Mo typically 1 - 2% Cr, ½% Mo)
P-5A	Low Alloy Steel (Cr-Mo typically 2 - 3% Cr, ½-1% Mo)
P-5B	Intermediate Alloy Steel (typically 5 - 10% Cr, 1% Mo)
P-5C	Low and intermediate alloy steel heat treated to 85 ksi or higher
P-6	Straight Cr, 400 Series, Martensitic Stainless Steel
P-7	Straight Cr, 400 Series, Ferritic Stainless Steel
P-8	Cr-Ni, 200 and 300 Series, Austenitic Stainless Steel
P-9A	up to 2½% Ni Steels
P-9B	3½% Ni Steels
P-9C	4½% Ni Steels
P-10A	Mn-V and Mn-½Ni-V Steels
P-10B	1Cr-V Steel
P-10C	C-Mn-Si Steel
P-10F	Mn-3Mo-V and 2Ni-2Cr-3Mo Steels
P-10G	36% Ni Steel
P-10H	Duplex Stainless Steels
P-10I	Ferritic Stainless Steels
P-10J	Ferritic Stainless Steels
P-10K	Ferritic Stainless Steels
P-11A	Heat Treated Low and Intermediate Alloy Steels
P-11B	Heat Treated Low and Intermediate Alloy Steels
P-21	Commercially Pure Aluminum & Al-Mn Alloys (AA1060, AA1100, AA3003)
P-22	Al-Mn & Al-Mg Alloys (AA3004, AA5052, AA5154, AA5254, AA5454, AA5652)
P-23	Al-Mg-Si Alloys (AA6061, AA6063)
P-25	Al-Mg Alloys (AA5083, AA5086, AA5456)
P-31	Cu & Cu Alloys
P-32	Admiralty Brass, Naval Brass, Aluminum Brass, Muntz Metal
P-33	Cu-Si Alloys
P-34	Cu-Ni Alloys
P-35	Al-Bronze Alloys
P-41	Ni Alloys and Commercially Pure Ni
P-42	Ni-Cu Alloys (Monels)
P-43	Ni-Cr and Ni-Cr-Mo Alloys (Inconels)
P-44	Mo-Cr-Fe Alloys (Hastelloys)
P-45	Fe-Ni-Cr-Mo-Cu (Incoloys)
P-46	Ni-Cr-Si Alloys
P-47	Ni-Cr-W-Co-Fe-Mo Alloys
P-51	Ti & Ti Alloys
P-52	Ti & Ti Alloys
P-53	Ti Alloys
P-61	Zr & Zr Alloys
P-62	Zr Alloys

Table 6.4 List of AWS Specifications for Welding Consumables

Spec.	Specification Title
A5.01	Filler Metal Procurement Guidelines
A5.1	Covered Carbon Steel Arc Welding Electrodes
A5.2	Carbon and Low Alloy Steel Rods for Oxyfuel Gas Welding
A5.3	Aluminum and Aluminum Alloy Electrodes for Shielded Metal Arc Welding
A5.4	Covered Corrosion-Resisting Chromium and Chromium-Nickel Steel Welding Electrodes
A5.5	Low Alloy Steel Covered Arc Welding Electrodes
A5.6	Copper and Copper Alloy Covered Electrodes
A5.7	Copper and Copper Alloy Bare Welding Rods and Electrodes
A5.8	Brazing Filler Metal
A5.9	Corrosion-Resisting Chromium and Chromium-Nickel Steel Bare and Composite Metal Cored and Stranded Welding Electrodes and Welding Rods
A5.10	Bare Aluminum and Aluminum Alloy Welding Electrodes and Rods
A5.11	Nickel and Nickel Alloy Welding Electrodes for Shielded Metal Arc Welding
A5.12	Tungsten Arc Welding Electrodes
A5.13	Solid Surfacing Welding Rods and Electrodes
A5.14	Nickel and Nickel Alloy Bare Welding Electrodes and Rods
A5.15	Welding Electrodes and Rods for Cast Iron
A5.16	Titanium and Titanium Alloy Welding Rods and Electrodes
A5.17	Carbon Steel Electrodes and Fluxes for Submerged Arc Welding
A5.18	Carbon Steel Filler Metals for Gas Shielded Arc Welding
A5.19	Specification For Magnesium Alloy Welding Electrodes And Rods
A5.20	Carbon Steel Electrodes for Flux Cored Arc Welding
A5.21	Composite Surfacing Welding Rods and Electrodes
A5.22	Flux Cored Corrosion-Resisting Chromium and Chromium-Nickel Steel Electrodes
A5.23	Low Alloy Steel Electrodes and Fluxes for Submerged Arc Welding
A5.24	Zirconium and Zirconium Alloy Welding Electrodes and Rods
A5.25	Consumables Used for Electroslag Welding of Carbon and High Strength Low Alloy Steels
A5.26	Consumables Used for Electrogas Welding of Carbon and High Strength Low Alloy Steels
A5.27	Copper and Copper Alloy Gas Welding Rods
A5.28	Low Alloy Steel Filler Metals
A5.29	Low Alloy Electrodes for Flux Cored Arc Welding
A5.30	Consumable Inserts

ASME F-Number (Filler Number) Classifications

All but one of the AWS specifications (A5.19 - Magnesium) have been adopted for use by ASME, so far without any changes. The numbers of the AWS specifications adopted by ASME are prefixed by the letters "SFA" to indicate acceptance by ASME.

ASME Section IX has further grouped electrodes and rods belonging to the detailed AWS specification and classification system, essentially according to the usability characteristics of the electrodes or rods, which fundamentally determines the ability of welders and welding operators to make satisfactory welds. F-Numbers (see Table 6.5) are assigned to each group with the stated intent of reducing the number of welding procedure and performance qualifications which must be carried out. Note that the alloy group associated with each F-Number is similar to the alloy group associated with the equivalent P-Number.

What this means in general is that the F-Number of the welding consumable stated on the welding procedure can not be changed without requalifying the procedure, but the AWS specification or classification within an F-Number group may be changed, subject of course to compliance with other essential variables (e.g., A-Number) and supplementary essential variables when required. However, ASME Section IX also indicates code groupings do not imply that filler metals within a group may be indiscriminately substituted for a filler metal used in the qualification test without consideration of the compatibility of the base and filler metals from the standpoint of metallurgical properties, postweld heat treatment, design requirements, mechanical properties, and service requirements.

Table 6.5 ASME F-Number List

F-No. Range	Material Group
F-No. 1 - F-No. 6	Steel and Steel Alloys
F-No. 21 - F-No. 24	Aluminum and Aluminum-Base Alloys
F-No. 31 - F-No. 37	Copper and Copper-Base Alloys
F-No. 41 - F-No. 45	Nickel and Nickel-Base Alloys
F-No. 51 - F-No. 54	Titanium and Titanium Alloys
F-No. 61	Zirconium and Zirconium Alloys
F-No. 71 - F-No. 72	Hard-Facing Weld Metal Overlay

A-Number (Analysis Number) Classifications for Ferrous Metals

In addition to F-Numbers for welding consumables, for ferrous metals ASME also has an A-Number system for classifying welding consumables according to deposited weld metal analysis (see Table 6.6). The A-Number can be interpreted as a control on the indiscriminate substitution of welding consumables based on F-Number alone.

As an example, for a procedure qualification carried out on a piece of NPS 6 Schedule 80, ASTM A 106 Grade B pipe with an F-No. 3 (E6010) root and F-No. 4 (E7018) weldout, on the basis of F-Number alone, it could be assumed that any F-No. 3/F-No. 4 combination of welding consumables has been qualified. However, when an A-Number is included in the evaluation, the range of qualified welding consumables is narrowed considerably on the basis of weld metal analysis. Filler metal manufacturers frequently list F-Numbers and A-Numbers in product brochures, and ¶QW404.5 lists methods of determining A-Numbers.

Table 6.6 ASME Analysis Number (A-Number) Classification

A No.	Type of Weld Deposit	Analysis, % [Note (1)] C	Cr	Mo	Ni	Mn	Si
1	Mild Steel	0.15	---	---	---	1.60	1.00
2	Carbon-Molybdenum	0.15	0.50	0.40-0.65	---	1.60	1.00
3	Chromium (0.4-2%)-Molybdenum	0.15	0.40-2.00	0.40-0.65	---	1.60	1.00
4	Chromium (2-6%)-Molybdenum	0.15	2.00-6.00	0.40-1.50	---	1.60	2.00
5	Chromium (6-10.5%)-Molybdenum	0.15	6.00-10.50	0.40-1.50	---	1.20	2.00
6	Chromium-Martensitic	0.15	11.00-15.00	0.70	---	2.00	1.00
7	Chromium-Ferritic	0.15	11.00-30.00	1.00	---	1.00	3.00
8	Chromium-Nickel	0.15	14.50-30.00	4.00	7.50-15.00	2.50	1.00
9	Chromium-Nickel	0.30	25.00-30.00	4.00	15.00-37.00	2.50	1.00
10	Nickel to 4%	0.15	---	0.55	0.80-4.00	1.70	1.00
11	Manganese-Molybdenum	0.17	---	0.25-0.75	0.85	1.25-2.25	1.00
12	Nickel-Chrome-Molybdenum	0.15	1.50	0.25-0.80	1.25-2.80	0.75-2.25	1.00
Note 1:	Single values shown above are maximums.						

Trade Names

The welding industry often uses trade names for welding consumables, and it can be difficult to obtain comparative AWS classifications. To assist in this exercise, *AWS Specification A5.0, Filler Metal Comparison Charts* is available and consists of about 300 pages of tables with trade names of welding consumables matched to their applicable AWS specification and classification. Similar information exists in other publications such as *Welding Engineering Data & Filler Metal Comparison Charts* published by the editors of *Welding Design & Fabrication* magazine. Finally, information is always available from consumable manufacturers, and they frequently publish filler metal comparison charts which position their products relative to other popular brands.

Positions [¶QW-405]

Pressure piping welds may be completed in the position in which they are found, or they may be completed by rolling the pipe and piping components about some axis of symmetry. Consequently, it is common to speak of "position welds" and "roll welds," respectively.

Most welding codes and standards have carefully defined welding positions and ASME Section IX is no exception. It illustrates positions for fillet welds and groove welds in plate and pipe [¶QW-461.4 and ¶QW-461.6] and provides limitations within the articles of ASME Section IX. For groove welds in piping butt joints, the following positions are typical:

a) In the 1G position, the pipe axis is in the horizontal plane, the pipe is rotated, and the weld is deposited at or near the top of the pipe, in the flat position.
b) In the 2G position, the pipe axis is in the vertical plane, and the weld is deposited while moving around the pipe in the horizontal plane.
c) In the 5G position, the pipe axis is in the horizontal plane, and the weld is deposited while moving around the pipe in the vertical plane.
d) In the 6G position, the pipe axis is inclined at a 45 degree angle, and the weld is deposited while moving around the pipe in vertical plane oriented at 90 degrees to the pipe axis.

Note that there are tolerances on positions and, for the purist, these are explained by diagrams in ¶QW-462.1 and ¶QW-462.2.

When a 5G or 6G position weld is deposited starting at the top of the pipe and working toward the bottom of each side of a pipe, the welding direction is called "vertical down." Conversely, when the weld is deposited starting at the bottom of the pipe and moving toward the top, the welding direction is called "vertical up." Welding direction has importance from three points of view:

a) training and experience of welders, since the techniques used to go uphill or downhill are different,
b) type of electrode, since the coating characteristics, resulting slag characteristics, and volume of weld metal influence capability for downhill welding, and
c) notch toughness, since the ability to stack weld metal during uphill welding can detrimentally affect impact properties.

Preheat and Interpass Temperature [¶QW-406 & ¶330]

In modern welding technology, it is common to speak of the weld thermal cycle and the effect that it has on the properties of base metals and weld metal. The total weld thermal cycle depends on:

a) energy input from the welding process (commonly known as heat input or arc energy input for arc welding processes, as discussed later in this section);
b) preheat temperature;
c) interpass temperature range; and
d) postweld heat treatment cycle, if PWHT is applied.

B31.3 defines *preheating* as the application of heat to the base metal immediately before or during a forming, welding, or cutting process [¶300.2]. This definition has a slightly broader scope than the definition provided in ASME Section IX ¶QW-492. B31.3 also states that preheat is used, along with heat treatment, to minimize the detrimental effects of high temperature and severe thermal gradients inherent in welding [¶330.1]. Although not specifically mentioned in B31.3, these detrimental effects can include:

a) cold (hydrogen) cracking,
b) hard, brittle heat affected zones,
c) distortion, and/or
d) high residual stress.

Preheat can also be used to assist in the fusion (melting) of metals with high conductivity such as copper and copper alloys, and thick sections of aluminum and aluminum alloys. Preheat is sometimes used to assist in the alignment of parts, but such practice is usually discouraged due to lack of control over the alignment activity and the potential introduction of abnormally high stresses.

ASME Section IX defines *preheat temperature* as the minimum temperature in the weld joint preparation immediately prior to welding; or in the case of multiple pass welds, the minimum temperature in the section of the previously deposited weld metal, immediately prior to welding [¶QW-492]. Note that the ASME definition of preheat temperature as applied to multipass welds is also known as *minimum interpass temperature* in other codes, standards, and specifications. In this book, the term *minimum interpass temperature* will be used when referring to the minimum temperature of the deposited weld metal before starting the next pass of a multipass weld since, from a technical perspective, it is not always necessary that the temperature between passes meet or exceed the minimum preheat temperature at the start of the first pass.

ASME Section IX defines *interpass temperature* as the highest temperature in the weld joint immediately prior to welding, or in the case of multipass welds, the highest temperature in the section of the previously deposited weld metal, immediately before the next pass is deposited [¶QW-492]. Again, strictly speaking, the interpass temperature as defined by ASME is the *maximum interpass temperature*. In practice, it is measured near the start of the next pass.

Most requirements for preheat and minimum interpass temperature control are aimed at prevention of cold cracking in transformation hardenable materials such as carbon steels, low alloy steels, intermediate alloy steels, and martensitic stainless steels. For cold cracks to form, the four conditions listed in the left column of Table 6-7 must be satisfied simultaneously. The magnitude of any one of the four conditions can not be defined with accuracy and depends largely on the other three conditions. The beneficial effect of preheat on each of these conditions is shown in the right column of Table 6.7.

Table 6.7 Influence of Preheat and Minimum Interpass Temperature on Conditions Required for Cold Cracking

Conditions Necessary for Cold Cracking	Benefits of Preheat and Minimum Interpass Temperature Control
Presence of hydrogen in sufficient quantity.	Diffuses hydrogen away from the joint, reducing the risk of cracking.
Sufficient tensile stress, which may be applied, residual, or a combination of both.	Alters stress distribution during welding and offers a small reduction of residual stress upon completion of welding.
A susceptible microstructure which is normally interpreted to mean a hard microstructure, hence the frequent use of hardness tests to assess the risk of cold cracking.	Reduces probability of forming susceptible microstructure by slowing the cooling rate.
A temperature threshold below some critical level (e.g., 300ºF), which depends somewhat on alloy content and metal structure.	Keeps the weldment above threshold temperature for cracking until the weld is completed.

Another term used by ASME Section IX is *preheat maintenance*. Although not specifically defined by the code, examination of ¶QW-406.2 shows that the term applies to the maintenance or reduction of preheat upon completion of welding, prior to any required postweld heat treatment. The primary reason for preheat maintenance after completion of welding is to continue diffusion of hydrogen from the weldment. It may also be used for reduction of thermal gradients and resulting residual stresses or distortions, and/or for isothermal transformation. When preheat maintenance is specified for a welding procedure, it should include both temperature and time. Simply indicating "yes" on the weld procedure specification is meaningless.

For most Code users, the following questions regarding preheat and interpass control must be answered:

a) What preheat and minimum interpass temperatures should be used for the application?
b) How should the heat be applied?
c) How, where, and when should the temperature be measured?

Detailed technical answers to the above questions can be very complex. Since most piping applications involve some legal requirement (contractual and/or jurisdictional), the obvious starting point for answers is to examine the minimum requirements of B31.3 and ASME Section IX.

Preheating for all types of piping welds is covered in B31.3 under ¶330. The second sentence of ¶330.1 states: "The necessity for preheating and the temperature to be used shall be specified in the engineering design and demonstrated by procedure qualification." If such requirements are not included in the engineering design, then the minimum Code requirements are generally assumed.

Table 330.1.1 gives minimum requirements and recommendations for preheat, based on P-Number and thickness. When selecting minimum preheats from this table, there are three additional conditions to consider:

a) For ambient temperatures below 32ºF (0ºC), recommendations of Table 330.1.1 become mandatory requirements [¶330.1.1]. Many owner specifications also contain a statement to the effect that code recommended preheat shall be interpreted as mandatory preheat.
b) When base metals of a dissimilar weld have different preheat requirements, the higher temperature shown in the table is recommended [¶330.2.3]. Again, many owner specifications may make this recommendation a mandatory requirement.
c) In the table, thickness refers to the thickness of the thicker component measured at the joint [¶330.1.1]. Strictly speaking, the joint is the interface between the members. In some cases, where end preparations have been placed on components, it may be necessary to project the original shape and dimensions of the component to determine governing thickness.

With respect to procedure qualification, it is assumed that the adequacy of minimum preheat and interpass temperatures is demonstrated by successful procedure qualification. If you can prepare a test weld using a given preheat and interpass temperature range, and the weld is satisfactory, then the preheat and interpass temperature range are deemed acceptable.

The demonstration concept is simple, but there are some pitfalls which must be addressed, especially for unusual design or construction circumstances. For example, most of the welding processes of ASME Section IX list ¶QW-406.1 as an essential variable. According to this clause, a decrease of up to 100ºF from the temperature used during procedure qualification may be permitted (subject, of course, to B31.3 ¶330.1.1 and Table 330.1.1 constraints). In terms of material response to the total weld thermal cycle, technical justification of ¶QW-406.1 becomes difficult, particularly in situations where hardness control of a weldment may be required. Note, however, that ASME Section IX does not concern itself with hardness control.

The situation with respect to ¶QW-406.1 is made more complex in view of the "2T" qualification rule of ¶QW-450. Basically, if a test weld is made on a coupon of thickness "T," this qualifies to weld on material of thickness "2T." Again, considering material response to the weld thermal cycle, how can one be assured of the adequacy of preheat at a temperature 100ºF less than used for the procedure qualification, when welds are completed on material with a thickness twice that of the test coupon? The simple answer is: you can't. However, it is good to recognize that the ASME Section IX rules have adequately supported the construction of many piping systems, pressure vessels, and boilers. The urge to supplement ASME requirements for perceived unusual design or construction circumstances must be balanced with the technical need for increased assurance of the preheat adequacy and interpass temperature control.

In addition to information contained in B31.3 Table 330.1.1, there are numerous other methods of evaluating or verifying requirements for preheat and interpass temperature control. Some techniques, with associated comments, are described below:

a) Nomographs are a favorite tool of technical specialists for estimating preheat requirements. One common method used for carbon and carbon-manganese steels was published by the British Welding Institute in a book entitled *Welding Steels Without Hydrogen Cracking* by F. R. Coe. The technique has a sound scientific basis in that it accounts for base metal thickness, arc energy input, carbon equivalent, and hydrogen potential of the welding process. The publication also describes techniques for dealing with alloy steels.

b) For the less technically inclined, the use of "look-up" tables is very common. These may be in very simple or very complex form. For ASME code users, the appendix of a book entitled *Weldability of Steels* by R. D. Stout and W. D. Doty can be a very useful source of preheat information. This appendix is based on WRC Bulletin 191 first published in January 1974 and republished in March 1978. Although the information may appear to be a bit dated (depending upon the current state of your career), it nevertheless addresses the complex issues of base metal thickness, carbon content and general chemistry (e.g., hardness and hardenability), and hydrogen potential.
c) Other sections of the ASME code are often consulted for preheat information including ASME Section VIII, Division 1, Appendix R; ASME Section VIII, Division 2, Appendix D; and ASME Section I, A-100.
d) Several types of preheat calculators have been developed over the years. One common calculator is available from the Lincoln Electric Company. It incorporates the influence of hydrogen, carbon equivalent, and thickness on preheat selection.
e) Complex preheat calculation techniques have been published, which relate cooling rates calculated by various formulae (e.g., solutions to Rosethal's equations for heat flow from a moving heat source) to time-temperature-transformation (TTT or equivalent) diagrams. However, these techniques find little application within the pressure piping industry, due to the complexity of calculations, the need to select values for "constants" which are not constant over the temperature ranges involved in welding, and the lack of complete time-temperature-transformation data for the materials involved. Of course, as computers continue to change our world, one may find increasing application of these more fundamental methods.

B31.3 does not restrict the methods of preheat, but some owner specifications do. Some methods of preheating include:

a) oxyfuel gas fired torch (propane, butane, and sometimes acetylene),
b) electric resistance elements,
c) induction coils, and
d) exothermic kits.

From a practical welding perspective, the method of heating is not generally an issue as long as the correct preheat temperature is achieved, with the heat uniformly applied throughout the full thickness and circumference of the joint. In some cases, the use of oxyacetylene torches is prohibited by owner specifications due to the intensity of the heat source and the risk of local damage to the base metal or weld metal if the heat is concentrated at one location.

B31.3 requires that the preheat temperature be checked to ensure that the temperature specified by the WPS is obtained prior to and during welding [¶330.1.3(a)]. However, B31.3 does not prescribe:

a) the exact methods by which temperature must be measured,
b) the location of temperature measurements, or
c) the timing of temperature measurements.

B31.3 only indicates that:

a) temperature indicating crayons, thermocouple pyrometers, or other suitable means shall be used to measure temperature, [¶330.1.3 (a)]
b) thermocouples may be attached by capacitor discharge welding without the need for welding procedure and performance qualifications, subject to visual examination of the area after thermocouple removal, [¶330.1.3 (b)] and
c) the preheat zone shall extend at least 1 inch beyond each edge of the weld. (Note that many owner specifications require at least 2 inches and some as much as 6 inches) [¶330.1.4].

Although the Code definitions of preheat and interpass temperature refer to the deposited weld metal, temperature measurement directly on hot weld metal can result in contamination and is generally discouraged. Common industry practice is to measure preheat and interpass temperature in a manner which insures that the correct preheat has been reached from the edge of the welding groove to the outer limits of the specified preheat zone width.

Preheat temperature, by definition, is measured immediately before the start of welding; however, interpass temperature measurement can be the subject of some debate. When interpass temperature control is required by a WPS, the interpass temperature is normally measured immediately before the start of the next pass, at the location where the next pass will be started.

Purists may argue that by only measuring the temperature at the starting point of the next pass, it is possible for the next pass to be placed over weld metal that is not exactly within the interpass range. It's true, of course, but practically speaking, interpass temperature measurement for process piping is hardly an exacting science.

Gas for Shielding, Backing, and Purging [¶QW-408]

Protection of hot and molten metal from the atmosphere is a necessary part of most welding operations. Protection may be accomplished using:

a) externally supplied shielding, backing, and purging gases, and/or
b) fluxes which decompose to give a slag covering and/or a gaseous shield.

Externally supplied protective gases prevent atmospheric contamination of the hot molten metal by displacing air from the weld area. In general, protective gases may be:

a) inert gases like helium or argon, which do not react with the hot metal,
b) reactive gases such as carbon dioxide, nitrogen, or hydrogen, which do react in a limited way with the hot metal and may be oxidizing or reducing depending upon the specific gas/metal interaction, or
c) mixtures of inert and/or reactive gases (e.g., $75Ar/25CO_2$).

For many welding applications, shielding gases are actually mixtures of gases, with the composition of the mixture optimized to provide the best combination of shielding characteristics and process operating characteristics. Selection of a gas depends on several somewhat interrelated factors including:

a) cost and availability at required purity levels,
b) ease of handling, stability, and physiological effects,
c) metallurgical characteristics including solubility in metals being welded, reactions with metals being welded and resultant degree of protection afforded to hot metal, effects on wetting behavior, effects on penetration as it affects type and thickness of metals to be welded, and effects on end properties of the weld deposit, and
d) welding process characteristics including welding position, ease of arc ignition as influenced by ionization potential of the gas, arc stability, and penetration as affected by thermal conductivity of the gas.

Solubility is a particularly important issue in gas selection, since dissolved gas in the molten metal can lead to porosity on freezing. Inert gases such as helium and argon have very limited solubility in most metals and are therefore used extensively for gas shielded arc welding processes. Although carbon dioxide is virtually insoluble in most metals, it is reactive and will cause some surface oxidation and some loss of oxidizable elements. Nevertheless, carbon dioxide is used extensively for GMAW and FCAW of carbon and low alloy steels.

In some cases, limited amounts of certain gases are included in the shielding gas mixture to accomplish objectives other than shielding. Three examples are listed below:

a) Nitrogen has been added to stainless steel shielding gases as an alloy addition which imparts improved corrosion resistance and strength, and as a technique of controlling phase balance in duplex stainless steel weld metal.
b) Oxygen may be added to argon in amounts typically from 1 to 5%, to improve wetting of stainless steel deposits, to improve bead shape, and to reduce undercutting.
c) Hydrogen has been added to shielding gases to increase penetration characteristics, through increased arc voltage and consequent increased heat input.

As a word of caution, minor additions of reactive gases should not be attempted without thorough understanding of the consequences, and procedure qualification to evaluate the effect of such additions.

Cleaning [¶328.4.1]

¶328.4.1 does provide some motherhood statements about cleaning, but a welding procedure should be specific about cleaning methods, solutions, abrasives, and tools. This is particularly true for nonferrous metals and stainless steels, since inappropriate cleaning methods can lead to cracking (prior to or in service) and/or loss of corrosion resistance.

Workmanship

¶328.5.1(d) prohibits peening on the root pass and final pass of a weld. Peening work hardens the metal, reduces ductility, and therefore increases the risk of cracking during or after welding. Peening between passes can be permitted because heat of subsequent weld passes heat treats (softens) the peened weld metal of previous passes. Note that chipping necessary for slag removal is not considered to be peening.

¶328.5.1(e) is intended to prohibit welding under adverse weather conditions. Moisture can cause porosity and hydrogen cold cracks. Excessive wind blows shielding away causing porosity and brittle welds, for both gas shielded and coated electrodes. Another practical issue with excessive wind under dry or arid conditions is the swirling of dirt within the welder's helmet, making it difficult to see the weld pool. If you can't see what you're doing, you can't weld.

¶328.5.1(f) provides some general advice on preserving the seat tightness of weld end valves. The valve manufacturer should always be consulted concerning welding conditions appropriate to maintaining seat tightness and responsibility for dismantling, reassembly, and testing when necessary. Some issues to consider include characteristics of sealing materials and risk of heat damage (e.g., plastic or metal-to-metal seat elements), use of extended bodies for soft seated valves, or use of a water quench to cool valve materials which do not transformation harden (e.g., austenitic stainless steels).

¶328.5.3 indicates that seal welds must be done by a qualified welder. While this requirement may seem obvious, significant commercial gains can be made through the use of unqualified welders. The need for seal welds to cover all threads is imposed to avoid notch effects which could cause brittle fracture or fatigue cracks. For the record, a seal weld is not a back weld.

Mechanical Testing

Mechanical testing requirements for welding procedures are found in ASME Section IX. However, if the base metal will not withstand the 180 degree guided bend test required by ASME Section IX, B31.3 permits qualification if the weld bend specimen will undergo the same degree of bending as the base metal (within 5 degrees) [¶328.2.1(b)].

As well, when impact testing is required by the Code or engineering design, B31.3 indicates that those requirements shall also be met in qualifying weld procedures [¶328.2.1(d)].

Heat Treatment [¶331]

Heat treatment is used to minimize certain detrimental effects associated with welding, bending, and forming processes [¶331]. Depending on the nature of each process, attendant high temperatures, severe thermal gradients, and/or severe metal forming operations (cold work) can result in dramatic loss of toughness, reduction of ductility, increased hardness, and/or high residual stresses. In turn, these can lead to premature, unexpected, and potentially catastrophic failures caused by brittle fracture, fatigue cracking, stress corrosion cracking, and/or hydrogen embrittlement.

B31.3 provides basic heat treatment practices suitable for most welding, bending, and forming operations, but warns that they are not necessarily appropriate for all service conditions [¶331]. Common examples where compliance with minimum B31.3 heat treatment requirements could be considered inappropriate include process streams containing caustics, amines used for gas sweetening operations, and hydrogen sulfide. In the case of process streams containing hydrogen sulfide, it is known that the minimum temperature allowed by Table 331.1.1 may not cause sufficient softening for resistance to sulfide stress cracking in severe sour environments. This accounts for the careful wording in clause 5.3.1.3 of NACE Standard MR0175-95, which states: "Low-alloy steel and martensitic stainless steel weldments shall be stress relieved at a minimum temperature of 620ºC (1150ºF) to produce a maximum hardness of 22 HRC maximum." Typically, temperatures required to satisfy the 22 HRC maximum criterion for low alloy steels are well above 620ºC (1150ºF). Given the toxic nature of sour environments, before construction one should confirm that proposed heat treatment cycles are capable of satisfying maximum hardness restrictions. Confirmation would typically involve cross-sectional hardness surveys of the welding procedure qualification test coupons.

Forms of Heat Treatment

There are many forms of heat treatment, each intended to accomplish a certain task. Frequently, more that one task is accomplished by a particular thermal cycle. Heat treatments listed in B31.3 Table 331.1.1 are best described as stress relieving heat treatments, since the primary purpose of the treatment is the reduction of residual stresses due to welding, forming, or bending operations. Such treatment may also result in improved ductility, lower hardness (note that B31.3 does impose some hardness restrictions), better toughness, and reduced distortion during subsequent machining operations. B31.3 also allows the use of annealing, normalizing, or normalizing and tempering, in lieu of a required heat treatment after welding, bending, or forming, provided that the mechanical properties of any affected weld and base metal meet specification requirements after such treatment and that the substitution is approved by the designer [¶331.2.1].

Stress relief of carbon and low alloy steel is carried out at a temperature slightly below the lower critical temperature (A_1) of the steel, hence there is some use of the term subcritical stress relief (i.e., there is no phase transformation at subcritical temperatures). For high alloy materials, such as austenitic stainless steels, considerably higher temperatures are required for effective stress relief due to the inherent hot strength of these materials. Although B31.3 does not mandate application of heat treatment to austenitic stainless steels, when applied (e.g., for service reasons), it is usually carried out at a temperature of approximately 1650ºF.

Annealing is a common term in the heat treating business, but there are many types of anneal. A full anneal is performed at high temperature (e.g., about 50 to 100ºF above the upper critical temperature for carbon steels), followed by slow cooling, generally in a furnace. It provides maximum softening, resulting in lowest hardness and strength. A stress relief anneal is performed on carbon and alloy steels at a temperature slightly below the lower critical temperature, and may also be known as a subcritical stress relief or subcritical anneal. Both the stress relief anneal and full anneal are applied as softening treatments, but other effects may result from the thermal cycles including changes to mechanical properties, physical properties, and microstructure. Although "full anneal" is generally implied when the word "anneal" is used without any qualifier, if a "full anneal" is required, fewer surprises will occur if the term "full anneal" is used.

Sometimes the term *solution anneal* is used to describe what might properly be called a *solution heat treatment*. In this case, an alloy is heated to a temperature high enough to dissolve one or more constituents into solid solution, and then cooled rapidly enough to hold the constituents in solid solution. Solution heat treatments are generally applied to high alloy steels and other high alloy materials for the purpose of dissolving one or more constituents which may affect the properties of the material. For example, many of the austenitic stainless steel pipes purchased according to ASTM A 312 are supplied in the solution treated condition.

Normalizing of carbon and low alloy steels is carried out by heating to a temperature range similar to that used for a full anneal, but the parts are allowed to cool in still air. This can save time and money compared to annealing, if the soft structure of the full anneal is not required. Normalizing is also effective in refining the grain size and homogenizing the structure, resulting in better toughness, more uniform mechanical properties, and better ductility.

Tempering is a heat treatment which may be applied to transformation hardenable steels after a normalizing operation and is generally applied after a quenching operation. Tempering is carried out below the lower critical temperature. It is used to reduce hardness and improve toughness and ductility at the expense of reduced strength.

Heat Treatment Requirements

¶331.1.1 imposes the following heat treatment requirements:

a) Heat treatment shall be in accordance with the material groupings and thickness ranges in Table 331.1.1 except as provided in ¶331.2.1 and ¶331.2.2.
b) Heat treatment to be used after production welding shall be specified in the WPS and shall be used in qualifying the welding procedure.
c) The engineering design shall specify the examination and/or other production quality controls (not less than the requirements of B31.3) to ensure that the final welds are of adequate quality.
d) Heat treatment for bending and forming shall be in accordance with ¶332.4.

Governing Thickness for Heat Treatment of Welds [¶331.1.3]

B31.3 contains a detailed treatment of metal thickness rules governing the need for heat treatment, as well as exceptions to the rules [¶331.1.3]. Table 6.8 below may be used to assist with interpretation of B31.3 requirements.

Table 6.8 Governing Thickness and Exemptions for
Postweld Heat Treatment (PWHT) of Welds

Weld Type	PWHT Governing Thickness And Exceptions
Butt Welds and welds not covered elsewhere in this table	PWHT required when the thickness of the thicker component measured at the joint exceeds the limits provided in Table 331.1.1. No exceptions.
Branch Connection Welds for set-on or set-in designs, with or without reinforcement, as per Fig. 328.5.4.D	PWHT required when the thickness through the weld in any plane is greater than twice the minimum material thickness requiring heat treatment as specified in Table 331.1.1. No exceptions. See B31.3 for assistance with weld thickness calculations.
Fillet Welds for slip-on, socket, and seal welded connections NPS 2 and smaller; and for external nonpressure parts such as lugs and pipe supports in all pipe sizes.	PWHT required when the thickness through the weld in any plane is more than twice the minimum material thickness requiring heat treatment as specified in Table 331.1.1. There are three exceptions to this rule: (1) Heat treatment is not required for P-No. 1 materials with weld throats 7⅝ in. (716 mm), regardless of base metal thickness. (2) Heat treatment is not required for P-No. 3, 4, 5, or 10A materials with weld throats 7½ in. (713 mm), regardless of base metal thickness, provided that preheat applied during welding was not less than the recommended preheat (see Table 330.1.1 and the WPS), and that base metal SMTS is less than 71 ksi (490 MPa). (3) Heat treatment is not required for ferritic materials when welds are made with filler metal which does not air harden. (See Note 1.)

(1) Note that this can be a dangerous exception for some services and should be used with caution. Just because the filler metals do not harden, doesn't mean the heat affected zones of the base metals have not hardened. Furthermore, just because the deposited filler metal may be in a soft and ductile condition, doesn't mean the base metal heat affected zones are in the same condition.

Equipment and Methods of Heat Treatment

B31.3 does not impose limitations on heating equipment and methods. It only indicates that the heating method must provide the required metal temperature, metal temperature uniformity, and temperature control, and then lists methods which may be used for heating including furnace, local flame heating, electric resistance, electric induction, and exothermic chemical reaction [¶331.1.4].

Heating methods used for heat treatment may be classified in terms of the facility used for heat treating and the energy source. Facilities can be discussed in terms of local heat treatment and furnace heat treatment.

a) Local heat treatment involves the heating of a small band of metal. Normally the band being heat treated is stationary, but in some manufacturing operations, the band moves. Examples of moving bands include in-line tempering operations used during manufacture of quenched and tempered pipe and local heating operations used in the manufacture of induction bends.

b) Furnace heat treatment generally involves placing the item to be heat treated inside a permanent furnace operated by a fabrication shop or commercial heat treater. However, it is possible to construct temporary heat treatment facilities (e.g., at the job site) which may range from simple ad hoc insulated box constructions to complex portable furnaces.

There are several energy sources used in heat treating. Current commercial sources of heat energy and characteristics are discussed below:

a) Electric resistance heat is produced when an electric current is passed through wires made from a material with high electrical resistivity. The electric current increases atomic movement in the wire, and this energy is then released in the form of heat. There is a wide selection of heating elements available in the marketplace, with various sizes and shapes to fit practically any geometry. Heaters are flexible and durable and are commonly used for local heat treatment of welds during field construction. Although resistance elements may burn out or short circuit against the pipe, there are several advantages supporting use of electric resistance heat, especially for local heat treatment of welds.
   i) Heat can be continuously and evenly applied.
   ii) Temperature can be adjusted accurately and quickly.
   iii) Welders can work in relative comfort. For preheating applications, they do not have to stop intermittently to raise preheat temperature.
   iv) Heat input can be adjusted fairly easily for example, to control the amount of heat applied to different pipe quadrants or sections of dissimilar thickness such as weld-in valves.
b) Heat can be produced by chemical reaction using exothermic kits. The chemical composition of constituents in modern exothermic heat kits is proprietary; however. when materials in the kit are reacted, heat is released. Heating cycles are controlled in terms of the size, shape, and heating value of the exothermic charge; the size, shape, and mass of the component to be heat treated; and the local environmental conditions. Although exothermic kits offer the advantages of portability, low capital equipment cost, and simple operator training, they have two major limitations:
   i) Once the kit has been ignited, it is difficult to perform any further adjustments.
   ii) It is usually difficult or impossible to satisfy Code and owner requirements on heating rate, holding time, and cooling rate.

   For these reasons, use of exothermic kits is restricted by many owner specifications, except perhaps for infrequent use at remote locations such as drilling sites. Successful application of exothermic kits requires careful planning and contingency measures.
c) Radiant heaters use infrared radiation generated by a gas flame or quartz lamp to develop heat. Infrared radiation is a form of electromagnetic radiation which behaves similarly to light. The intensity of the radiation falls off in proportion to the square of the distance between the heat emitter and the part heated. Radiation reaching the part is either absorbed, causing the temperature of the part to increase, or reflected (wasted). Consequently, surface condition of the metal will affect the efficiency of the process, as will the relative positions of the heater and the part, since the heater must "see" the part for effective heating.
d) Induction coils create heat with the passage of alternating current (ac). The alternating magnetic field associated with the alternating electrical field penetrates the metal to be heated,

changes strength and direction in phase with the external alternating electric current, and produces eddy currents in the part to be heat treated. The rise and collapse of magnetic fields and their associated eddy currents stimulate atomic movement resulting in the release of heat within the part.

Temperature Measurement

B31.3 does not impose restrictions on the devices used for temperature verification [¶331.1.6]. Although devices such as temperature indicating crayons, thermometers, and optical pyrometers may be used, thermocouple pyrometers are generally used to measure and record the temperature of the surfaces being heat treated. For proper temperature measurement, the hot end of the thermocouple junction must be in direct contact with the surface of the pipe or kept at the same temperature as the pipe by being inserted into a terminal joined to the pipe.

B31.3 does not impose any restrictions on thermocouple attachment methods. To this end, one may observe on various job sites, attachment of thermocouples using steel bands, wire, ad hoc clamping devices, or weld metal. The suitability of any of these thermocouple attachment methods should be addressed by specifications developed during the engineering design or by QA/QC personnel monitoring the work. For example, steel bands or wires can become loose at the heat treatment temperature so that the thermocouple is no longer in contact with the surface, leading to false temperature readings. Ad hoc clamping devices can work loose, and the attachment method of such devices can cause local damage to the pieces being heat treated. Use of weld metal to fix the thermocouple to the pipe surface changes the composition of the junction, resulting in measurement error. Experience with attachment of thermocouples by capacitor discharge welding indicates that the process works well. B31.3 has specific statements permitting the attachment of thermocouples to pipe using capacitor discharge welding without the need for weld procedure and performance qualifications [¶331.1.6, ¶330.1.3(b)].

B31.3 does not address the number of thermocouples required or the placement of thermocouples. Although these issues should be dealt with in construction specifications, perhaps discussion of a few sources of measurement error would be useful in making such determinations.

a) Heat rises. With local heat treatment, the temperature measured on the top of a pipe will normally be higher than the temperature measured on the bottom of the pipe. Typically, as diameter increases, steps are necessary to allow introduction of more heat at the bottom and sides of the pipe than at the top of the pipe. The need for additional temperature measurement points and heating controls does have commercial implications.
b) The pipe will be hottest adjacent to the source of the heat. During local heat treatment, the pipe is normally heated from the outside, so the inside surface will be a bit cooler than the outside surface, depending upon the thickness of the pipe and the extent to which insulation is used and drafts prevented. For example, if the bore of the pipe is not plugged with insulation and the wind is whistling through it, a substantial difference between the inside and outside surface temperatures might be expected. Since heat treatments are often conducted for service reasons,

and the service is usually on the inside of the pipe, specific attention is required to ensure that the inside surface is adequately heated (and that the outside surface is not excessively heated).

c) The surface temperature of a resistance coil or other radiant heat source is considerably above that of the pipe being heated. If the hot junction of the thermocouple is not insulated from the heat source, the temperature reading will be higher than the actual temperature at the pipe surface.
d) Thermocouple wires should run along the pipe surface under the insulation for several inches before coming outside the insulation on the pipe surface. If the wires are brought straight out of the insulation from the point of contact with the pipe, heat may be conducted away from the hot junction leading to a temperature reading lower than the pipe surface.
e) Measurement errors may be introduced if extension wires are not of the same composition as the thermocouple wire, all the way from the hot junction to the cold junction. Don't accidentally reverse the wires at a connection point.
f) Ensure that instrumentation is properly calibrated. Battery operated circuits should be calibrated at regular intervals and the output of regulated power supplies should be checked occasionally for accuracy.
g) Damaged or contaminated thermocouples and extensions can lead to measurement error and should be checked regularly for physical damage (severe bends, kinks, partially broken wires, weld spatter or slag trapped between the wires).

Heating and Cooling Rates [¶331.1.4]

B31.3 does not impose restrictions on heating and cooling rates [¶331.1.4]. ASME Section VIII, Division 1, UCS-56 requirements are frequently applied, but there are cases where insulation and coils are ripped off immediately after the soak period. In the stress relieving business, time is money, especially if the heat treater is working by the weld or lump sum. So, if heating and cooling rates need to be controlled, it should be stated in specifications or other contract documents.

Hardness Tests [¶331.1.7]

B31.3 makes the following statements regarding hardness testing.

> Hardness tests of production welds and of hot bent and hot formed piping are intended to verify satisfactory heat treatment. Hardness limits apply to the weld and to the heat affected zone (HAZ) tested as close as practicable to the edge of the weld.
>
> a) Where a hardness limit is specified in Table 331.1.1, at least 10% of welds, hot bends, and hot formed components in each furnace heat treated batch and 100% of those locally heat treated shall be tested.

b) When dissimilar metals are joined by welding, the hardness limits specified for the base and welding materials in Table 331.1.1 shall be met for each material.

The Code does not discuss many of the technical details necessary to give an accurate and representative appraisal of production weld hardness. Consequently, owner specifications are recommended for guidance on applying this simple but often misused and abused test method. Owner specifications should consider the size of hardness indentations relative to the size of weld zones to be measured, surface preparation of the weld, methods for locating the zones of interest, and training requirements for hardness testing personnel.

CHAPTER 7

INSPECTION, EXAMINATION, AND TESTING

Introduction

Inspection, examination, and testing are activities carried out to ensure that piping systems meet the minimum requirements of the B31.3 Code and the engineering design. Clauses governing these activities are found mainly in B31.3 Chapter VI (Inspection, Examination, and Testing), with additional requirements in Chapter VIII (Piping for Category M Fluid Service) and Chapter IX (High Pressure Piping).

Inspection Versus Examination

Under B31.3 rules of construction, it should be noted that inspection and examination do not mean the same thing. Table 7.1 compares certain defining characteristics for each activity. Although there may appear to be similarities between the actual work of inspectors and examiners on the job, it is important to be able to distinguish the responsibilities associated with each activity. Our ability to make such distinction has improved with the expansion of quality assurance and quality control concepts within the piping industry.

Table 7.1 Interpretive Comparison of Inspection and Examination

	Inspection [¶340.1, ¶340.2]	**Examination** [¶341.1, ¶341.2]
Corporate Responsibility:	Owner.	Manufacturer, Fabricator, or Erector.
Individual Responsibility:	Owner's inspector or delegates of the owner's inspector.	Examination (QC) personnel.
Work Description:	Verify that all required examinations and tests have been completed. Inspect piping to the extent necessary to be satisfied that it conforms to all applicable examination requirements of the Code and the engineering design.	Perform examinations required by B31.3. (Note that most QC manuals have sections devoted specifically to completion of examinations, such as material control, welding control, NDE control, pressure testing, and record keeping.)
Primary Quality Management Function:	Quality assurance, including quality audit.	Quality control.

Personnel Requirements [¶341]

Qualifications for owner inspectors and examination personnel are covered in ¶340.4 and ¶342, respectively. Table 7.2 compares inspector and examiner qualifications according to certain defining characteristics.

Table 7.2 Comparison of Requirements for Inspectors and Examiners

	Owner's Inspectors [¶340.4]	**Examination Personnel** [¶342.1, ¶342.2]
Appointment	Inspectors shall be designated by the owner, and shall be the owner, an employee of the owner, an employee of an engineering or scientific organization, or of a recognized insurance or inspection company acting as the owner's agent.	B31.3 does not list any specific requirements. Examiners are usually employees of the manufacturer, fabricator, or erector, or employees of a service agency subcontracted by the manufacturer, fabricator, or erector.
Restrictions	Inspectors shall not represent or be an employee of the piping manufacturer, fabricator, or erector, unless the owner is also the manufacturer, fabricator, or erector.	In-process examinations must be performed by personnel other than those performing the production work [¶342.2].
Education and Experience	B31.3 requires that inspectors have 10 or more years experience in the design, fabrication, or inspection of industrial piping (3). However, each 20% of satisfactorily completed work toward an engineering degree recognized by the Accreditation Board for Engineering and Technology can be considered equivalent to 1 year of experience, up to 5 years total [¶340.4(b)].	B31.3 is very loose regarding personnel qualification and certification. It simply states that "examiners shall have training and experience commensurate with the needs of the specified examinations." By a reference note to [¶342.1], for evaluation of personnel, B31.3 indicates that SNT-TC-1A may be used as a guide. (2)
Certification	No requirements stated.	B31.3 requires that the employer certify records of examiners employed, showing dates and results of personnel qualifications, and maintain the records and make them available to the inspector.

(1) The term *inspector*, when used according to the B31.3 Code, does not necessarily mean an Authorized Inspector as written in other sections of the ASME code.
(2) While SNT-TC-1A is certainly a useful document, if such qualifications or equivalent are required, it must be specifically stated in contract documents. Furthermore, the level of qualification (I, II, or III) should be stated in terms of requirements for specific parts of the work. As an example, for radiographic exposure of circumferential butt welds in a process piping system, an ASNT Level I radiographic qualification should be adequate. For interpretation of the resulting radiographs, an ASNT Level II radiographic qualification should be specified. In cases of interpretation dispute, access to an ASNT Level III qualified radiographic examiner is desirable.
(3) B31.3 does not prescribe the methods used to ensure that inspectors satisfy the requirement. Some organizations have implemented testing systems to assess the qualifications of inspectors and examiners. Most tests are written and can reflect the knowledge base of the candidate, but can not necessarily address the field experience factor. Therefore a good deal of judgment is required in hiring owner inspectors.

Examination [¶341]

For most B31.3 Code users, examination requirements can be summed up by the following questions:

a) What items must be examined?
b) What types of examinations must be applied to the items?
c) When must the items be examined?
d) What extent of examination is required?
e) How should the examinations be conducted?
f) What are the standards of acceptance applicable to each examination?
g) What disposition should be assigned to nonconforming items?

The above questions look simple enough, but finding the answers can be time consuming and frustrating, especially since examination terminology is foreign, inconsistent, and confusing to many users. When assessing examination requirements for a project (i.e., answering the above questions), the usual starting point is a review of B31.3 and contract clauses, followed by a listing of examination requirements, and then by developing tables, if appropriate. Several tables have been included in this chapter to illustrate B31.3 examination requirements. Note that these tables express the author's interpretation of B31.3 requirements. They should be carefully reviewed and supplemented where necessary prior to project use.

What Items Must Be Examined?

Most examination requirements are applicable to welds, but examinations may also be necessary for other items including castings and bends. Items requiring examination depend upon fluid service. The left columns of Tables 7.4 and 7.5 list several items which may require examination depending upon service classification.

What Types of Examinations Must Be Applied to the Items?

B31.3 lists seven *types of examination*:

1. visual examination [¶344.2]
2. magnetic particle examination [¶344.3]
3. liquid penetrant examination [¶344.4]
4. radiographic examination [¶344.5]
5. ultrasonic examination [¶344.6]
6. in-process examination [¶344.7]
7. progressive examination [¶341.3.4]

The first five types of examination are also referred to as *methods of examination*, which can be confirmed by referring to ASME Section V, Article 1. Progressive examination is included as a type of examination, even though it is only used when defects are revealed by spot or random examination.

An *NDE (nondestructive examination) method* is generally described in terms of the probing medium used to detect surface and internal discontinuities in materials, welds, and fabricated parts and components. Examples include the radiographic method, which uses electromagnetic radiation as a probing medium, and the ultrasonic method, which uses high frequency sound waves (ultrasound) as a probing medium (see Table 7.3).

An *NDE technique* is a specific way of using a particular NDE method [ASME Section V, Article 1]. For example, a weld could be examined by the ultrasonic method using a shear wave technique, where the sound beam propagates in a shear wave mode.

An *NDE procedure* is an orderly sequence of actions describing how a specific technique shall be applied [ASME Section V, Article 1].

Table 7.3 NDE Methods, Abbreviations, and Probing Mediums

NDE Method	Abbreviation	Probing Medium
Visual Method	VT	Visible light
Magnetic Particle Method	MT	Magnetic field
Liquid Penetrant Method	PT	Liquid
Radiographic Method	RT	Electromagnetic radiation
Ultrasonic Method	UT	High frequency sound waves
Note: The "T" used in the above abbreviations represents "test" and is derived from terminology used by the American Society for Nondestructive Testing (ASNT).		

When Must the Items Be Examined?

B31.3 provides the following information regarding the timing of examinations:

a) Required examinations must be performed prior to initial operation [¶341.3.1 and ¶K341.3.1].
b) For P-Nos. 3, 4, and 5 materials, examinations must be performed after completion of any heat treatment [¶341.3.1(a)]. The primary reason for this requirement is the increased risk of reheat cracking associated with these low alloy steels. For any high pressure piping material subject to heat treatment, ¶K341.3.1 requires that examination be conducted after completion of heat treatment.
c) For welded branch connections, examinations and repairs must be completed before addition of any reinforcing pad or saddle [¶341.3.1(b)]. This requirement is mainly of a practical nature and aims to avoid any work constraints imposed by the restricted access or interference created by repads or saddles.

What Extent of Examination is Required?

Determination of the extent of examination required actually involves answering two questions:

a) How many items must be examined by the method (e.g., lot extent)?
b) How much of each item must be examined by the method (e.g., item extent)?

Number of Items Requiring Examination (Lot Extent)

The number of items to be examined by any particular method of examination is defined in terms of a percentage of the total lot (e.g., 5%, 100%). It depends first on the category of fluid service, then on the kind of item to be examined (e.g., weld, type of weld, type of component, threads), and finally on the type of examination. Some Code clauses dealing with the extent of examination are ¶341.4, ¶M341.4, and ¶K341.4.

For most fluid services, B31.3 clauses distinguish between "Visual Examinations" and "Other Examinations" which are more sophisticated forms of nondestructive examination. Tables 7.4 and 7.5 provide an overview of the extent of "visual examination" and "other examination" required for various types of fluid service and kinds of items to be examined.

Table 7.4 Extent of Visual Examination Required by B31.3

Items to be Visually Examined	**Category D Fluid Service** [¶341.4.2]	**Normal Fluid Service** [¶341.4.1(a)]	**Severe Cyclic Conditions** [¶341.4.3(a)]	**Category M Fluid Service** [¶M341.4]	**High Pressure K Service** [¶K341.4.1]
Materials and Components	(2)	Sufficient Random Selection	Sufficient Random Selection	Sufficient Random Selection	100% (5)
Fabrication	(2)	5%	100%	100%	100%
Longitudinal Welds (1)	(2)	100%	100%	100%	100%
Joints (Mechanical)	(2)	Sufficient Random Selection (3)	100%	100%	Sufficient Random Selection (4)
During Erection	(2)	Random	Random	Random	Random
After Erection	(2)	Required, but amount is not specified	All (see ¶341.4.3(a)(3))	Required, but amount is not specified	Required, but amount is not specified

(1) Refers to longitudinal welds made during fabrication.
(2) Visual examination only required to the extent necessary to satisfy the examiner of B31.3 conformance.
(3) Sufficient to ensure conformance with ¶335, except that all joints shall be examined for piping subject to pneumatic test.
(4) Sufficient to ensure conformance with ¶335, except that all joints shall be examined for piping subject to pneumatic test, plus 100% examination of pressure containing threads.
(5) Also see ¶K302.3.3 regarding casting examination.

Table 7.5 Extent of Other Examination Required by B31.3

Items to be Examined	**Category D Fluid Service** [¶341.4.2]	**Normal Fluid Service** [¶341.4.1(b)]	**Severe Cyclic Conditions** [¶341.4.3(b)]	**Category M Fluid Service** [¶M341.4(b)]	**High Pressure K Service** [¶K341.4.2]
Circumferential Butt and Miter Groove Welds	None	min. 5% random RT or UT (1, 2)	100% RT (2,3)	min. 20% random RT or UT (1, 2, 3)	100% RT of all girth welds (6, 7)
Longitudinal Welds	None	RT or UT intersections (9)	(10)	RT or UT intersections (9)	100% RT (6, 7, 8)
Branch Welds	None	None	100% RT of fabricated branches (4)	min. 20% random RT or UT (1, 2, 3)	100% RT (6, 7)
Fabricated Laps	None	None	None (11)	min. 20% random RT or UT (1, 2, 3)	None
Socket Welds	None	None	100% MT or PT (5)	None	None
Brazed Joints	None	min. 5% In-Process Examination	None	None	None

(1) Welds shall be selected so that work of each welder or welding operator is included in the examination [¶341.4.1(b)(1)].

(2) In-process examination may be substituted on a weld-for-weld basis if specified in the engineering design or specifically authorized by the inspector [¶341.4.1(b)(1)].

(3) In-process examination shall be supplemented by appropriate NDE [¶341.4.3(c), ¶M341.4(2)].

(4) Refers to fabricated branch welds which are 100% radiographable. 100% UT may be used in lieu of RT, if specified in engineering design. Socket welds and branch connection welds which are not radiographed shall be examined by magnetic particle or liquid penetrant methods [¶341.4.3(b)].

(5) Includes branch connection welds not subject to RT [¶341.4.3(b)].

(6) UT shall not be substituted for RT, but UT can supplement RT [¶K341.4.2(b)].

(7) In-process examination shall not be substituted for RT [¶K341.4.2(c)].

(8) Also see ¶K305.1.1 and ¶K305.1.2 for additional information.

(9) For the required random radiography of circumferential butt and miter groove welds, shot locations shall be selected to maximize coverage of intersections with longitudinal joints. At least 1 1/2" (38 mm) or the longitudinal welds shall be examined [¶341.4.1(b)(1)].

(10) ¶341.4.3(b) does not specify other examination of longitudinal welds. However, ¶305.2.3 in Chapter II of B31.3 limits the pipe which may be use under severe cyclic conditions. To obtain the required joint efficiencies for welded pipe, some RT of the longitudinal seam may be required depending upon specifications used to purchase the pipe. The amount of required RT may be assessed by consulting B31.3 Table A-1B to obtain the joint efficiency of the pipe as procured. If this joint efficiency does not satisfy ¶305.2.3, then additional RT must be completed according to B31.3 Table 302.3.4 to raise the joint efficiency to the required level.

(11) ¶341.4.3(b) does not make specific reference to examination of fabricated laps. For additional welding and examination reference regarding the use of fabricated laps for severe cyclic conditions see ¶306.4.3 and ¶311.2.2.

Amount of Examination Applied to Each Item (Item Extent)

To determine how much examination must be applied to each item an understanding of the following terminology as defined by ¶344.1.3 is required:

a) *100% examination* is complete examination of all of a specified kind of item in a designated lot of piping.
b) *Random examination* is complete examination of a percentage of a specified kind of item in a designated lot of piping.
c) *Spot examination* is a specified partial examination (e.g., percentage) of each of a specified kind of item in a designated lot of piping.
d) *Random spot examination* is specified partial examination of a percentage of a specified kind of item in a designated lot of piping.

For radiographic examination, the amount of examination is further refined by ¶344.5.2, as follows:

a) *100% radiography* applies only to girth and miter groove welds and to fabricated branch connections comparable to Fig. 328.5.4E (i.e., radiographable branches), unless otherwise stated in the engineering design.
b) *Random radiography* applies only to girth and miter groove welds.
c) *Spot radiography* requires a single exposure radiograph in accordance with ¶344.5.1 at a point within the specified extent of welding. Additional details are provided in ¶344.5.2.

In a) and b) above, the amount of radiography is essentially limited in terms of the capability to perform radiographic examination (i.e., radiographable connections). Item c) above supplements the general spot examination definition by providing specific information on the size of spot which must be examined when spot radiography is required.

With the above terminology in mind, extent of required examination may be determined from Code paragraphs including those summarized in Tables 7.4 and 7.5.

How Should Examinations Be Conducted?

B31.3 addresses how examinations should be conducted under heading ¶344. This is summarized by Table 7.6 according to examination method and type of item to be examined (castings and welds or components other than castings). Note that the articles of ASME Section V provide rules governing the conduct of a test, but the articles can not be construed as procedures.

What Are the Standards of Acceptance?

Standards of acceptance for B31.3 pressure piping depend upon fluid service, type of examination applied, and type of item examined. Table 7.7 is a summary of references to standards of acceptance for welds based on fluid service and type of examination.

Table 7.6 References for Examination Procedures

Examination Type/Method	B31.3 Clause	Welds (1)	Castings
Visual	¶344.2.2	ASME Sect. V, Art. 9	ASME Sect. V, Art. 9
Magnetic Particle	¶344.3	ASME Sect. V, Art. 7	¶302.3.3
Liquid Penetrant	¶344.4	ASME Sect. V, Art. 6	¶302.3.3
Radiographic	¶344.5.1	ASME Sect. V, Art. 2	¶302.3.3
Ultrasonic	¶344.6	ASME Sect, V, Art. 5 (2)	¶302.3.3
In-Process	¶344.7	ASME Sect. V, Art. 9 (3)	ASME Sect. V, Art. 9 (2)

(1) Column for welds includes components other than castings, with the exception of ultrasonic examination, where components other than castings are not covered.
(2) See ¶344.6.1 for alternates to ASME Sect. V, Art. 5, Para. T-543.1.3 and T-547.1.1.
(3) Additional methods may be specified in the engineering design.

Table 7.7 References to Standards of Acceptance for Welds

Examination Type/Method	Category D Service	Normal Service	Severe Cyclic Service	Category M Service (1)	High Pressure Service
Visual (VT) ¶341.3.2	¶341.4.2 Table 341.3.2A - Category D Fluid Service	¶341.4.1 Table 341.3.2A - Normal Fluid Service	¶341.4.3 Table 341.3.2A - Severe Cyclic Conditions	¶M341, M341.4 Table 341.3.2A - Normal	¶K341.3.2 Table K341.3.2A
Radiographic (RT) ¶341.3.2	N/A (2)	¶341.4.1 Table 341.3.2A - Normal Fluid Service	¶341.4.3 Table 341.3.2A - Severe Cyclic Conditions	¶M341, ¶M341.4 Table 341.3.2A - Normal	¶K341.3.2 & Table K341.3.2A
Ultrasonic (UT) ¶341.3.2	N/A (2)	¶344.6.2	¶344.6.2	¶M341, ¶M341.4, ¶M344, ¶344.6.2	¶K341.3.2, ¶K341.5, ¶344.6.2 (3, 4)
Magnetic Particle (MT) ¶341.3.2	N/A (2)	N/A (5)	¶341.4.3 Table 341.3.2A - Severe Cyclic Conditions	N/A (5)	N/A (5)
Liquid Penetrant (PT) ¶341.3.2	N/A (2)	N/A (5)	¶341.4.3 Table 341.3.2A - Severe Cyclic Conditions	N/A (5)	N/A (5)

(1) Note that many designs for toxic and lethal fluids impose standards of acceptance above the minimum requirements of B31.3.
(2) Such examination is not mandatory for Category D service.
(3) Note that ultrasonic examination may not be substituted for radiography, but may supplement it [¶K341.4.2(c)].
(4) Supplementary examination requirements for pipe and tubing, above that required by the material specifications, are covered by ¶K344.6.2 through ¶K344.6.4, and ¶K344.8.
(5) MT and PT are not required examinations for welds in this service. If MT and PT are specified in the engineering design, the standards of acceptance must also be specified.

B31.3 does not address all examination requirements and associated standards of acceptance encountered in daily practice. A review of Code interpretations published over the last decade shows several replies which simply indicate that B31.3 does not address the issue of concern, or that the issue of concern is only partially addressed. Some examples are listed below.

a) Interpretation 12-22 asks whether the allowable cumulative length of a defect would be prorated downward for welds with lengths less than the lengths shown by the criterion value notes B, C, F, and G in Table 341.3.2A. As an example, if the criterion value limit was a cumulative length of 1.5 in. in any 6 in. of weld length, what would be the maximum cumulative length of defect in a weld 4 in. long. The reply simply states: "The Code does not specifically address this situation."
b) Interpretation 8-38 indicates that B31.3 does not address radiography of pipe welds for Category D fluid service. Although the owner may specify radiography of these welds in the engineering design, the acceptance criteria must also be stated since the acceptance criteria in Table 341.3.2A apply to visual examination.
c) Interpretation 8-32 asks whether one would use the size of indication or size of discontinuity as the basis for acceptance or rejection when using MT or PT techniques. The reply states "B31.3 specifies liquid penetrant and magnetic particle examination for detection of cracks only, and all cracks are rejectable." One must therefore conclude that there are no criteria governing rounded indications or other linear indications. If these other indications must be evaluated, the engineering design must specify the acceptance criteria.

A continuing difficulty with B31.3 is the confusion created by incorporation of NDE methods and acceptance criteria for welds within the same tables (i.e., Table 341.3.2A and Table K341.3.2A). For example, for normal fluid service, visual examination is indicated as a required examination for undercutting, but radiography is not a required examination for undercutting. If undercutting were observed on the radiographic film of a girth weld for normal fluid service, theoretically it would not be a radiographic call. (Note that, in practice, many radiographers regularly evaluate film for undercut, whether required by code or not.) One reason why undercut is not evaluated by radiography for normal fluid service, is the manner in which acceptance criteria are specified. Criterion values "H" and "I" limit the depth of undercut, which is difficult to establish using radiographic methods. In contrast, for severe cyclic conditions, undercutting is a radiographic call. The criterion value "A" prohibits any undercut so if undercut is visible on the film, it is rejectable. There is no need to evaluate undercut depth on radiographs for severe cyclic service.

Testing

B31.3 requires that each piping system be leak tested to ensure tightness [¶345.1] after completion of any required heat treatment [¶345.2.2(b)] and prior to initial operation [¶345.1]. The Code imposes certain restrictions on the types of leak tests which may be used [¶345.1], some general requirements for conducting leak tests [¶345.2], some preparation requirements for leak tests, some specific requirements for each type of leak test [¶345.4 through ¶345.9], and some record keeping requirements [¶345.2.7, ¶346]. However, the Code does not address every issue for every kind of leak test, and it is generally necessary to make a leak test plan incorporating Code requirements and non-Code requirements.

The following six types of leak test are described in B31.3:

a) Hydrostatic Leak Test [¶345.4]
b) Pneumatic Leak Test [¶345.5]
c) Hydrostatic-Pneumatic Leak Test [¶345.6]
d) Initial Service Leak Test [¶345.7]
e) Sensitive Leak Test [¶345.8]
f) Alternative Leak Test [¶345.9]

Table 7.8 compares and contrasts several characteristics of the six leak tests listed in B31.3. Note that the sensitive leak test is not applied as a stand alone test. Rather, it is used to fulfill part of the requirement for an alternative leak test.

Although six types of leak test are listed in the Code, a hydrostatic leak test is normally required [¶345.1], unless the following conditions are satisfied:

a) An initial service test may be used for a Category D fluid system at the owner's option [¶345.1(a)]. Typical candidates for initial service test include plant air systems and cooling water systems.
b) A pneumatic or hydrostatic-pneumatic leak test may be used where the owner considers a hydrostatic leak test impractical and the hazards associated with use of a compressible fluid are recognized [¶345.1(b)].
c) An alternative leak test may be used where the owner considers both hydrostatic and pneumatic leak tests to be impractical [¶345.1 (c)] and the conditions of ¶345.1(c)(1) and ¶345.1(c)(2) are satisfied.

Some general requirements for leak tests are outlined below. For a complete description of each requirement, readers should refer to the current version of B31.3.

a) If the pressure used during a leak test will produce a stress exceeding the yield stress at test temperature, the test pressure may be reduced to the maximum pressure that will not exceed the yield strength at the test temperature [¶345.2.1(a), ¶345.4.2(c)].
b) Precautions shall be taken to avoid excessive pressure due to thermal expansion of the test fluid [¶345.2.1(b)].
c) A preliminary pneumatic test with air at a maximum pressure of 25 psig may be used to locate major leaks [¶345.2.1(c)]. Note that some owners frown on this practice on the basis that use of the test may lead to sloppy construction work.
d) Leak tests shall be maintained for at least 10 minutes, and all joints and connections shall be examined for leaks [¶345.2.2(a)]. This means that the hold time may easily exceed 10 minutes, so that all joints and connections may be examined.
e) Leak tests shall be conducted after any heat treatment has been completed [¶345.2.2(b)]. This reduces the risk of brittle fracture due to low ductility microstructures existing prior to heat

treatment and increases the probability of detection of defects formed during heat treatment (e.g., reheat cracks).

f) B31.3 requires that the possibility of brittle fracture be considered when leak tests are conducted near the ductile-brittle transition temperature [¶345.2.2(c)]. Unfortunately, B31.3 does not define the term ductile-brittle transition temperature in a practical way, and transition curves from which the transition temperature may be obtained are rarely available. In lieu of such data, lowest test temperature could be considered as the lowest temperature for which impact tests would be required according to B31.3 rules.

g) B31.3 allows piping subassemblies to be tested separately or as assembled piping [¶345.2.3(a)]. This means, for example, that it is not mandatory to leak test piping spools in the fabrication shop and then again in the field. Of course, some owners require both shop and field leak tests to ensure that the shop fabricator has fulfilled contractual requirements, and they are prepared to pay the cost of testing.

h) For practical reasons, a flanged joint at which a blank has been inserted to isolate other equipment during a test need not be tested [¶345.2.3(b)].

In preparing for a leak test, B31.3 imposes the following requirements, which are only a few of the elements of a leak testing plan:

a) All joints, including welds and bonds, are to be left uninsulated and exposed for examination during leak testing, except that joints previously tested according to B31.3 may be insulated or covered. For sensitive leak tests, all joints shall be unprimed and unpainted [¶345.3.1].

b) Temporary supports may be necessary to accommodate the weight of the test fluid used for leak testing of vapor or gas lines [¶345.3.2].

c) Expansion joints represent a special testing situation. Refer to B31.3 for additional information [¶345.3.3, ¶345.4.2(c), Appendix X, Para. X3.2.3(a)].

d) At the limits of tested piping, equipment which is not included in the test shall be disconnected from the piping or isolated by blinds or other means. Valves may be used at the limits of tested piping, provided that the valve is suitable for the test pressure [¶345.3.4].

Table 7.8 Summary of Leak Tests and Their Characteristics

	Hydrostatic Leak Test	**Pneumatic Leak Test**	**Hydrostatic-Pneumatic Leak Test**	**Initial Service Leak Test (5, 6)**	**Sensitive Leak Test**	**Alternative Leak Test**
Test Fluid	Water or other suitable nontoxic liquid (1, 2)	Air or other nonflammable nontoxic gas	Water (1,2) and Air or other nonflammable nontoxic gas (12)	Service fluid (may be liquid, gas, or vapor)	See ASME Section V, Article 10	None (7)
Temperature	Selection of test temperatures should consider risk and consequences of brittle fracture [¶345.2.2(c), ¶345.5.1].					(13)
Pressure	$\frac{1.5 \times P \times S_T}{S}$ (3, 4)	110% P (3)	110% P (3, 10)	Operating Pressure	Lesser of 15 psi or 25% P (3)	None (7)
Hold Time	10 minutes minimum hold time required for all leak tests [¶345.2.2(a)].					N/A
Examination	All joints and connections must be examined for leaks [¶345.2.2(a)] (8).					100% NDE (9)
Pressure Relief Devices	Consider [¶345.2.1(b)]	Required [¶345.5.2]	Required [¶345.6 & ¶345.5.2] (11)	Consider [¶345.2.1(b)]	Consider [¶345.2.1(b)]	Not applicable
References To Other Code Requirements & Precautions	Testing through pressure equipment [¶345.4.3]	Stored energy & risk of brittle fracture [¶345.5.1] Rate of pressure increase [¶345.5.5] Preliminary checks [¶345.5.5]	Stored energy & risk of brittle fracture [¶345.5.1]	Rate of pressure increase [¶345.7.2] Preliminary checks if fluid is gas or vapor [¶345.7.3]	Test sensitivity [¶345.8] Rate of pressure increase [¶345.8(b)]	Flexibility Analysis [¶345.9.2] Sensitive Leak Test [¶345.9.3]

(1) Another suitable nontoxic liquid may be used if there is a possibility of damage due to freezing of water or due to the adverse affects of water on the piping or process. If the liquid is flammable, flash point shall be at least 120°F (49°C), and consideration shall be given to the test environment [¶345.4.1].
(2) In cold climates, water with a freezing point depressant such as glycol is often used for hydrotesting.
(3) P = Internal Design Gauge Pressure, S_T = stress value at test temperature, and S = stress value at design temperature. When S_T and S are equal, test pressure is 1.5 x P.
(4) For design temperature above the test temperature, see ¶345.4.2(b).
(5) Service leak test is also known as a commodity test.
(6) Use of service leak test requires owner acceptance [¶345.7 and ¶345.1(a)].
(7) A sensitive leak test is required as part of the alternative leak test [¶345.9.3].
(8) For initial service leak tests, ¶345.7.3 allows one to omit examination of any joints or connections previously tested according to the Code. For example, if part of the piping system subject to a service leak test was previously leak tested by another Code specified method, examination of that part of the piping system could be omitted.
(9) In addition to the examination requirements for sensitive leak tests, during alternative leak tests, welds not subjected to hydrostatic or pneumatic leak tests in accordance with B31.3, including those welds used in the manufacture of welded pipe and fittings, shall be examined as follows. Circumferential, longitudinal, and spiral groove welds shall be 100% radiographed. All other welds, including structural attachment welds, shall be examined by PT or MT (for magnetic materials).
(10) Pressure in the liquid filled part of the piping system shall not exceed limits stated in ¶345.4.2.
(11) Set pressure shall not be higher than the test pressure plus the lesser of 50 psi (340 kPa) or 10% of the test pressure [¶345.5.2].
(12) Although not stated in the Code, gas and liquid should be non-reactive.
(13) Test temperature should be compatible with selected NDE methods.

B31.3 makes the piping designer, manufacturer, fabricator, and erector responsible, as applicable, for preparing test records required by the Code and the engineering design [¶346.2]. Test records required by B31.3 include:

a) date of test [¶345.2.7(a)],
b) identification of piping system tested [¶345.2.7(b)],
c) test fluid [¶345.2.7(c)],
d) test pressure [¶345.2.7(d)],
e) certification of results by the examiner [¶345.2.7(e)],
f) examination procedures [¶346.3(a)], and
g) examination personnel qualifications [¶346.3(b)].

Items a) through e) need not be retained after completion of the test if the inspector certifies that the piping has satisfactorily passed a pressure test as required by the Code [¶345.2.7]. Items f) and g) shall be retained for at least 5 years after the record is generated for the project [¶346.3].

Chapter 8

PIPING FOR CATEGORY M FLUID SERVICE

Introduction

B31.3 establishes design, fabrication, inspection, and material requirements for piping systems designated by the owner as Category M Fluid Service in Chapter VIII. That chapter, with specific imposed requirements from the base Code (the first six chapters of B31.3) and Chapter VII (requirements for nonmetallic piping), [¶M300] is a stand-alone Code for piping classified by the plant owner as Category M Fluid Service. These requirements as well as the definition and Code responsibilities for Category M Fluid Service are the subject of this chapter.

Definition

The definition of a *Category M Fluid Service* is a fluid service in which the potential for personnel exposure to toxic fluids is judged to be significant. Exposure by breathing or bodily contact to even a very small quantity of such fluid, caused by leakage, can produce serious irreversible harm to persons even when prompt restorative measures are taken. [¶300.2(b)]

Classification Responsibility

In addition to the responsibility for overall compliance with B31.3 [¶300 (b)(1)], the owner is responsible for determining if a fluid service is Category M. B31.3, in Appendix M, provides a logic flow chart to assist the plant owner in determining which is the appropriate fluid service for the system. The considerations included are:

1. Is the fluid toxic?
2. Does the definition of Category M Fluid Service also describe the fluid in question?
3. Does the base Code (first seven chapters of B31.3) sufficiently protect personnel from exposure to very small quantities of the fluid in the environment?
4. Can the occurrence of severe cyclic conditions be prevented by design?

If the answers to 1, 2, and 4 are "yes" and the answer to 3 is "no", the fluid service for the particular piping system is Category M. The rules for the design of such Category M Fluid Service systems are found in Chapter VIII of B31.3 and are discussed in the following section.

Design Conditions

Design Temperature and Pressure

For the purpose of component wall thickness, component pressure rating, material allowable stress, and fatigue analysis, the design temperature shall be based on the fluid temperature. Design temperature method of determination other than fluid temperature is permitted provided this temperature is established by heat transfer calculation procedures consistent with industry practice. [¶M301.3]

For the purpose of component wall thickness and pressure rating, the design pressure is determined by the same procedure as the base Code. Design pressure is the most severe pressure expected in a service, coincident with temperature, which results in the greatest component pressure rating and the greatest component wall thickness. [¶M302.1]

Pressure-temperature variations permitted in ¶304.2.4 of B31.3 are not permitted in Category M Fluid Service. [¶M302.2.4]

Design Considerations

Two design considerations are specifically emphasized for special consideration in Category M Fluid Service piping. These are *impact* [¶M301.5.1] and *vibration* [¶M301.5.4]. Impact caused by water hammer or the equivalent of steam hammer should be eliminated as much as possible with piping layout and valve selection (particularly check valves). Where impact or vibration are recognized as unavoidable in plant start-up, shut-down, or normal operation, then pipe restraints, snubbers, and controls shall be employed to eliminate detrimental effects on the piping and restraints. The location and type of pipe restraint shall be determined by a dynamic analysis computer simulation of the vibrating piping system.

Wind and earthquake analysis per the procedures of ASCE 7-88 (soon to be revised to ASCE 7-93) is required for Category M Fluid Service piping as for base Code piping. These topics and methods of analysis are presented in Chapter 3 - Wind Loads and Earthquake.

Allowable Stresses/Allowances for Pressure Design

The basis of the allowable stress of metallic materials at temperature, is the same as that for materials used in the base Code. Even the use of materials not listed in Table A-1, where the allowable stresses are presented, is permitted provided the designer fully documents the basis of allowable stress determination and demonstrates that the procedure of determination is consistent with the procedure in ¶302.3.2 of B31.3. [¶M302.3]

Pressure Design of Metallic Piping Components

Pipe Wall Thickness for Internal Pressure

Pressure design for piping and piping components for Category M Fluid Service uses the base Code procedure in ¶304 of B31.3. [¶M304] The wall thickness determination for internal pressure systems is per equations (3a), (3b), or (3c) where the designer elects to employ outside diameter equations or by equation (3d) for calculations based on the pipe inside diameter. An example of the wall thickness calculation for internal pressure, outlined in Chapter 2, page 23. Wall Thickness for Internal Pressure, is as follows:

1. Calculate the required wall thickness, t, to contain the design pressure at temperature by equations (3a), (3b), (3c), or (3d) of B31.3.
2. Add the mill under run tolerance, corrosion/erosion allowance, thread depth or groove depth.
3. Select the next commercially available nominal wall thickness (schedule).

Pipe Wall Thickness for External Pressure

External pressure design, covered in ¶304.1.3 of B31.3 and discussed with examples of application in Chapter 2, page 27. Wall Thickness for External Pressure, is appropriate for determining the wall thickness of external pipe and components under external pressure including vacuum service. This wall thickness calculation procedure is presented in paragraphs UG-28 through UG-30 of the ASME Code, Section VIII Division 1 as specified by B31.3. Briefly, this is an iterative procedure where an initial trial, commercially available, wall thickness is chosen for the vacuum service pipe and the selected wall thickness is tested per the UG-28 (paragraph 2.2.2) procedure.

Limitations on Metallic Pipe, Pipe Fittings, and Bends

Metallic Pipe Material

Metallic pipe material listed in Table A-1 may be used in Category M Fluid Service with the exception of pipe specifically restricted to Category D Fluid Service [¶305.2.1] and piping requiring safeguarding [¶305.2.2]. [¶M305.2]

Metallic Pipe Fittings

Metallic pipe fittings manufactured in accordance with listed standards contained in Table 326.1 and unlisted pipe fittings qualified to B31.3 ¶302.2.3 may be used in Category M Fluid Service with the exception that the following shall not be used: [¶M306]

a) fittings conforming to MSS SP-43,
b) proprietary Type C lap-joint stub-end butt welding fittings. [¶M 306.1].

Pipe bends not manufactured in accordance with listed standards (Table 326.1) may be used in Category M Fluid Service piping provided (1) the bends are free of cracks, (2) the wall thickness of the extrados is not less than the required pressure design wall thickness as determined by B31.3 equations 3a, 3b, 3c, or 3d with corrosion, erosion, and mechanical allowances added, and (3) the out-of-round of the bend does not exceed 8% for internal pressure and 3% for external pressure limits established in ¶332.2.1 of B31.3. The temperature of the material during bending shall be per ¶332.2.2 of B31.3, (cold bending of ferritic materials shall be at a temperature below the transformation range; hot bending at a temperature above the transformation range).

Corrugated and creased bends as well as miter bends that make a change in direction at a single joint with an angle greater than 22.5° (see B31.3 Figure 304.2.3) shall not be used in Category M Fluid Service piping.

Branch Connections

For Category M Fluid Service, the pressure design of branch intersections that employ branch fittings or geometries not manufactured in accordance with standards listed in Table 326.1 shall be designed in accordance with base Code ¶304.3. These unlisted intersection components, such as unreinforced fabricated branch connections, or pad reinforced branch connections, are designed by the area replacement method presented in ¶304.3 and are discussed with examples in Chapter 2 - Branch Connections, page 33.

General Restrictions on Metallic Valves and Specialty Components

The following sections address most of the restrictions and modifications to the base Code rules for these components.

Valves [¶M307]

1. Valves with threaded bonnet joints, other than union joints, shall not be used.
2. Bonnet or cover plate closures shall be flanged, secured by at least four bolts and shall include appropriate gasketing, or they shall be secured by full penetration welds, or by straight thread using metal-to-metal seat and seal weld.

Flanges [¶M308]

Flanges qualified for use in base Code piping (piping designed in accordance with the first six chapters of B31.3) are suitable for use in Category M Fluid Service except as follows:

1. Single welded slip-on flanges shall not be used.
2. Expanded-joint flanges shall not be used.
3. Slip-on flanges used as lapped-joint flanges shall not be used, except for slip-on flange intersection of the flange face with the bore modified with a bevel or a radiused edge of approximately ⅛ in. (3 mm).
4. Threaded metallic flanges shall not be used, except those using lens rings or similar gaskets and those used in lined pipe where the liner extends over the gasket face.

Split Backing Rings

Welded joints with split backing rings are not allowed.

Socket Welded Joints

Socket weld joints greater than NPS 2 are not permitted.

Expanded Joints

Expanded joints are not allowed.

Flexibility and Support of Metallic Piping

Flexibility of Category M Fluid Service Piping

The flexibility rules of the base Code, as presented in Chapter 3 of this book, are applicable for Category M Fluid Service piping with the exception of equation (16), $Dy \div (L - U)^2 \leq K1$, used to determine the need for a formal analysis. This simplified procedure is not allowed in Category M Fluid Service. [¶M319]

Pipe Supports

The pipe support rules of the base Code (Section 321) apply to Category M Fluid Service piping except that steel of unknown specifications shall not be used as pipe supports. [¶M321]

Pressure Relieving Systems

The base Code rules for over-pressure protection apply to Category M Fluid Service piping with one exception. The set pressure on pressure relieving devices shall be set so the piping internal pressure will not exceed 110% of the design pressure in the pressure relieving state. [¶M322.6]

Metallic Piping Materials

Listed and unlisted materials that conform to published specifications and comply with the rules of the base Code, and reclaimed materials that have been properly identified and inspected can be used in Category M Fluid Service piping. Materials of unknown specifications and cast iron shall not be used in pressure containing parts. Lead and tin shall only be used as linings. [¶M323]

The use of listed carbon steel materials in low temperature Category M Fluid Service (below -20°F but above -50°F) is permitted, but impact testing is required. This is a deviation from the base Code where for these materials impact testing is not required. Other restrictions on the service conditions of carbon steels used in this low temperature range are the same for both fluid services, Category M and base Code. Hoop stress due to internal pressure shall not exceed 25% of allowable stress at ambient temperature, S_h, for cold temperature service, and the combined stresses caused by pressure, weight, and thermal displacement shall not exceed 6 ksi ($S_L + S_E < 6$ ksi). It is interesting to note that for the first time, the Code has established a maximum allowable stress for an operating condition (thermal displacement, pressure and weight combined). [¶M323.2]

Fabrication and Erection of Category M Fluid Service Piping

The rules for fabrication of Category M Fluid Service piping are the same as for base Code piping with the following exceptions.

Backing Rings [¶M328.3]

Split backing rings are prohibited, and removable backing rings and consumable inserts may be used only where their suitability has been demonstrated by procedural qualifications.

Pipe Bends

Creased or corrugated bends shall not be used. [¶M332]

Inspection, Examination, and Testing of Metallic M Fluid Service Piping

Inspection

The requirements for base Code piping apply to Category M Fluid Service piping with the following exceptions. [¶M340]

A. Visual Examination
   1. All fabrication shall be examined. [¶341.4]
   2. All threaded, bolted, and other mechanical joints shall be examined.

B. Other Required Examination

At least 20% of all, circumferential butt and miter welds, and fabricated lap and branch connection welds comparable to B31.3 Figure 328.5.4E and 328.5.5 sketches (d) and (e), shall be examined by either radiography or ultrasonic examination.

In-process examination, as allowed for base Code examination, is also allowed for Category M Fluid Service pipe examination if specified in the engineering design or by the inspector. In-process examination, if applied, shall be supplemented with appropriate nondestructive examination.

Leak Testing

Base Code rules apply to leak testing Category M Fluid Service piping with one exception. Sensitive leak testing per ¶345.8 shall also be included as a part of the hydrostatic or pneumatic testing. [¶M345]

Chapter 9

HIGH PRESSURE PIPING

Scope and Definition

The B31.3 rules for high pressure piping design are alternatives that become the basis of design only when the plant owner designates the piping as high pressure fluid service. When a plant owner designates a piping system to be high pressure fluid service, all the provisions of B31.3, Chapter IX become mandatory.

The plant owner is assisted in making the decision whether or not impose high pressure requirements by three simple guidelines.

1. If the design pressure of a particular piping system is higher than that which an ASME B16.5 Class 2500 flange can safely contain, then high pressure piping rules are required. For example, for ASTM A 105 material at 100°F, design pressures greater than 6,170 psig would require piping systems to be designed in accordance with the high pressure rules.

2. The maximum allowable stress, S_h, for carbon steel and alloy steel, for example, at elevated temperatures, is based on ⅔ yield at temperature. This limit on S_h for ferritic steels will not allow a B31.3 design temperature to exceed about 600°F. Heat treated austenitic stainless steels, (whose S_h value in base Code service is permitted to be as high as 90% of material yield strength at temperature), will be limited to a maximum temperature of about 800°F. There are no provisions for allowing S_h to be based on creep properties of any material. The S_c and S_h values are tabulated in Appendix K, Table K-1 of B31.3.

3. High pressure piping rules are not applicable to Category M Fluid Service.

With these three design conditions known, the owner will have sufficient information for deciding whether or not to impose high pressure piping requirements.

When the plant owner designates a piping system to be high pressure, all the requirements, Chapter IX of B31.3 becomes mandatory. Chapter IX becomes a stand-alone Code, drawing requirements from the first six chapters of B31.3 and modifying these provisions as appropriate for high pressure piping. The remainder of this chapter will focus on most of these modified requirements.

Modified Base Code Requirements for High Pressure Piping

Responsibilities of the Designer

The designer is responsible to the owner for compliance with the Code for all engineering design. In high pressure piping, this Code compliance shall be presented in the form of a written report, summarizing the results of the design analysis and the designer shall certify compliance with the B31.3 Chapter IX rules.

Design Conditions

Design Pressure and Temperature

The design pressure shall be based on the highest pressure the piping system will experience. The use of allowances for pressure variations as described in the base Code ¶302.2.4 is not permitted. [¶K301.2.1]

The design temperature shall be based on the fluid temperature. The presence or the absence of thermal insulation has no bearing on this temperature determination.

Pressure Design of Piping Components

Wall Thickness for Straight Pipe under Internal Pressure

The most significant deviation from the base Code is in the equations for determining the wall thickness requirements for internal pressure. Two equations are presented - one is based on the specified outside diameter (equation 34a), and one is based on the specified inside diameter (equation 34b). [¶K304.1.2] These equations are based on the Von Mises theory of failure.

Equation (34a), when used with S_h = ⅔ yield strength of the material, will produce a wall thickness with a pressure safety factor of at least 2. An example of the application of this equation follows:

What is the required wall thickness for pressure design for an NPS 12 EFW pipe constructed of ASTM A 106 Gr. B material with a design temperature of 300°F and a design pressure of 8,000 psig? The corrosion/erosion allowance is 0.063 in. The equation is:

$$t_m = t + c$$

$$t = \frac{D}{2}\left[1 - e^{\left(-1.155\frac{P}{S}\right)}\right]$$

where D = 12.75 in., P = 8000 psig, and S = 20,700 psi (from Table K-1, Appendix K).

Note: Even though an EFW pipe construction is specified, high pressure piping requirements are that the longitudinal weld joint factor E be fully examined producing an E factor = 1.0.

Solution:

$$t = \frac{12.75}{2}\left[1 - e^{\left(\frac{-1.155 \times 8000}{20{,}700}\right)}\right]$$

$t = 2.295$ in.

then $t_m = 2.295 + 0.063 = 2.358$ inches

Recognizing that this wall thickness is greater than the maximum scheduled pipe wall for this pipe size, this pipe will be custom made. The designer must determine the mill under-run tolerance for the manufacturing process and add that tolerance to the t_m value before making the purchase.

B31.3 also offers an ID equation for high pressure piping where the designer can calculate the pressure design wall thickness requirement for ID controlled-minimum wall thickness pipe. This equation (equation 34b) is:

$$t = \frac{d + 2c}{2}\left[e^{\left(\frac{1.155P}{S}\right)} - 1\right]$$

where d = pipe inside diameter. All other terms are as defined earlier.

An example of the application of this equation follows.

Find the pressure design wall thickness, t, for the same pipe considered above with the same design conditions:

d = 8.160 in.; P = 8,000 psig; c = 0.063 in.; T = 300°F; S = 20,700 psi

Solution: $$t = \frac{8.16 + 2 \times 0.063}{2}\left[e^{\left(\frac{1.155 \times 8000}{20{,}700}\right)} - 1\right]$$

$t = 2.331$ in.

In addition to equations for calculating wall thickness, the Code also presents two equations for calculating the maximum allowable internal design gage pressure, P. The first equation for pressure determination is based on the outside diameter, and the second equation is based on the inside diameter. The following presents an example of each, using the same conditions from each of the above examples

(D = 12.75 inches, and T = 2.358 inches for the outside diameter calculation and d = 8.16 inches and T = 2.331 inches for the inside diameter example):

The O.D. equation:

$$P = \frac{S}{1.155} Ln\left[\frac{D}{D - 2(T - c)}\right]$$

$$P = \frac{20{,}700}{1.155} Ln\left[\frac{12.75}{12.75 - 2(2.358 - 0.063)}\right]$$

$$P = 8{,}000 \text{ psig}$$

The ID equation for P:

$$P = \frac{S}{1.155} Ln\frac{(d + 2T)}{(d + 2c)}$$

$$P = \frac{20{,}700}{1.155} Ln\left(\frac{8.16 + 2 \times 2.331}{8.16 + 2 \times 0.063}\right)$$

$$P = 7{,}825 \text{ psig}$$

Wall Thickness for Straight Pipe Under External Pressure

The procedure for determining the pressure design wall thickness for external pressure depends upon on the pipe D/t ratio, where t is calculated by using the appropriate OD or ID based equation (34a or 34b) presented above. For the condition $D/t < 3.33$ and at least one end of the pipe is exposed to full external pressure by the presence of a weld cap, for example, that produces a compressive axial stress in the pipe wall, then the wall thickness for external pressure is to be calculated by the same equations for internal pressure (Code equations (34a) or (34b)). [¶K304.1.3]

The procedure for calculating the wall thickness for external pressure in high pressure piping where there is no axial compressive stress due to the presence of a component such as a weld cap, for example, is the same as that of the B31.3 base Code piping. The rules of ASME Section VIII Division 1 paragraphs UG-28 through UG-30 shall be followed except that the allowable stress values shall be taken from Table K-1.

Pipe Bends

The high pressure rules regarding wall thickness reduction at pipe bends are the same as the rules in the base Code; the wall after bending shall not be less than t_m. The major difference imposed for high pressure piping is that the radius of the pipe bend shall not be less than ten times the nominal outside diameter of the pipe. [¶K304.2] The same base Code rules for out-of-roundness apply to high pressure bends (3% for external pressure and 8% for internal pressure); however, bending temperature for quenched and tempered ferritic steels has additional limitations. The temperature at bending for cold bent ferritic steels shall be at least 50°F below the tempering temperature. [¶K332.2] Post bending heat treatment is required for hot bent piping materials with P-Numbers 3, 4, 5, 6, 10A, and 10B that are not quenched and tempered. [¶K332.4]

Miter bends are not allowed in high pressure piping. [¶K304.3.2]

Branch Connections

The strength of branch connections not manufactured in accordance with the listed standards of Table K326.1 shall be in accordance with the base Code rules for extruded outlets. [¶K304.3.2] These are area replacement rules, discussed and illustrated in Chapter 2 - Extruded Outlet Header, p38. Fabricated unreinforced or pad reinforced branch connections are not permitted. [¶K304.3.3] Proof testing in accordance with ¶K304.7.2 is an alternative method for qualifying unlisted intersections for high pressure. [¶K304.3.2]

Design of Other Piping Components for High Pressure Piping

The pressure design of other high pressure piping components, closures, flanges, blanks, blind flanges, and reducers that are not manufactured in accordance with standards listed in Table K326.1 must conform to base Code requirements which invoke the design procedures of the ASME Boiler and Pressure Vessel Code. These procedures were discussed in earlier chapters of this book. [¶K304.5]

Flexibility and Fatigue Analysis of High Pressure Piping

A thermal flexibility analysis of all piping systems shall be performed, and the resulting S_E as calculated by the base Code procedures shall not exceed S_A. The equation for calculating the allowable displacement stress range S_A, is the same as that provided by the base Code except that the stress range reduction factor, f, is always equal to 1.0 since displacement cycles greater than 7,000 are not allowed.

The calculated sustained load stress, S_L, shall not exceed S_h; this is not changed from the base Code. The combined sustained load with occasional load analysis is also required; however, the allowable stress is 1.2 S_h as compared to this same analysis allowable, 1.33 S_h of the base Code.

A fatigue analysis is required for all piping systems and all components within the system. The effects of all structural pipe supporting attachments shall also be investigated. This fatigue analysis shall be in accordance with ASME Section VIII Division 2, and the analysis shall include the effects of pressure and temperature. The amplitude of the pressure and temperature alternating stresses are to be determined in accordance with Division 2, Appendix 4 and 5. The allowable amplitude for the calculated alternating stresses shall be determined from the applicable fatigue curves of Division 2, Appendix 5.

Pressure Stress Evaluation for Straight Pipe Fatigue Analysis

For fatigue analysis, the stress intensity on the inside surface of straight pipe can be calculated using the equation: [¶K304.8.4]

$$S = \frac{PD^2}{2(T-c)\left[D-(T-c)\right]}$$

An example of the application of this pressure intensity equation is as follows:

Calculate the inside surface stress intensity of a 14 in. outside diameter pipe, T = 3.50 in., c = 0.125 in., and an internal pressure of 8,000 psig.

Solution:

$$S = \frac{8{,}000 \times 14^2}{2(3.5-0.125)\left[14-(3.5-0.125)\right]}$$

$$S = 21{,}863 \text{ psi}$$

If this calculated intensity exceeds three times the allowable stress from Table K-1 (Appendix K) at the average temperature during the loading cycle, then an inelastic analysis is required. [¶K304.8.4]

Appendix

1

AWS Classification System

AWS.	Specification Title, Classification Examples, and Explanation
A5.1	Covered Carbon Steel Arc Welding Electrodes - Example E7018-1 1 E designates an electrode. 2 The first two digits, in this case 70, indicate the minimum tensile strength of the deposited metal in the as-welded condition. For E7018-1, the minimum tensile strength is 70 ksi (70000 psi). 3 The third digit, in this case "1", indicates the position in which satisfactory welds can be made. a. "1" means the electrode is capable of satisfactory welds in all positions (ie. flat, vertical, overhead, & horizontal). b. "2" indicates the electrode is only suited to flat position welding and to horizontal position welding of fillet welds. c. "4" indicates the electrode is suitable for vertical-down welding and for other positions as described in AWS A5.1. 4 The last two digits taken together indicate the type current with which the electrode can be used, and the type of covering on the electrode. In the E7018-1 example, the number "8" identifies the electrode as suitable for AC or DC operation, with a predominantly lime (calcuim carbonate) coating. For other examples, see AWS Spec A5.1.
A5.2	Carbon and Low Alloy Steel Rods for Oxyfuel Gas Welding - Example R60 1 The letter R at the beginning of each classification designation stands for rod. 2 The digits (45, 60, 65, and 100) designate a minimum tensile strength of the weld metal, in the nearest thousands of pounds per square inch, deposited in accordance with the test assembly preparation section of this specification.
A5.3	Aluminum and Aluminum Alloy Electrodes for Shielded Metal Arc Welding - Example E1100 1 The letter E at the beginning of each classification designation stands for electrode. 2 The numerical portion of the designation in this specification corresponds to the Aluminum Association composition of the core wire used in the electrode. In the case of E1100, it is commercially pure aluminum.

AWS	Specification Title, Classification Examples, and Explanation (Continued)
A5.4	Covered Corrosion-Resisting Chromium and Chromium-Nickel Steel Welding Electrodes - Example E309LMo-16 1 The letter E at the beginning of each number indicates an electrode. 2 The first three digits of the classification indicate composition. In a few cases the number of digits may vary but composition is still indicated. 3 Letters may follow the digits to indicate specific alloy additions. In the case of the example, the letters "L" and "Mo" refer to a low carbon grade with a 2.0 to 3.0% molybdenum addition. 4 The last two digits of the classification indicate use with respect to position of welding and type of current. The smaller sizes of electrodes (up to and including 5/32 in. [4.0 mm]) included in this specification are used in all welding positions.
A5.5	Low Alloy Steel Covered Arc Welding Electrodes - Example E8018-B2L 1 The letter E designates an electrode 2 The first two digits (or three digits of a five digit number) designate the minimum tensile strength of the deposited metal in 1000 psi. For example E8018-B2L has a minimum tensile strength of 80000 psi. 3 The third digit (or fourth digit of a five digit number) indicates the position in which satisfactory welds can be made with the electrode. a. "1" means the electrode is satisfactory for use in all positions (flat, vertical, overhead, & horizontal). b. "2" indicates the electrode is suitable for the flat position and for horizontal fillet welds. 4 The last two digits, taken together, indicate the type current for the electrode and the type of covering on the electrode. 5 A letter suffix, such as A1, designates the chemical composition of the deposited weld metal.
A5.6	Copper and Copper Alloy Covered Electrodes - Example ECuNi 1. The letter E at the beginning of each number indicates a covered electrode 2. The chemical symbol Cu identifies the electrode as a copper-base alloy 3. Additional chemical symbols such as the Ni in ECuNi indicate the principal alloying elements of each classification or classification group. 4. If more than one classification is included for an alloy group, individual classifications are identified by the suffix letters A, B, C, etc., as in ECuSn-A. 5. Further subdivision within an alloy group is achieved using a numeral after the suffix letter (eg. the 2 in ECuAl-A2.

AWS	Specification Title, Classification Examples, and Explanation (Continued)
A5.7	Copper and Copper Alloy Bare Welding Rods and Electrodes - Example ERCuNi 1. The letters ER at the beginning of a classification indicate that the bare filler metal may be used either as an electrode or as a welding rod. 2. The chemical symbol Cu is used to identify the filler metals as copper base alloys. 3. Additional chemical symbols such as the Ni in ERCuNi indicate the principal alloying elements of each classification or classification group. 4. If more than one classification is included for an alloy group, individual classifications are identified by the suffix letters A, B, C, etc., as in ERCuSn-A. 5. Further subdivision within an alloy group is achieved using a numeral after the suffix letter (eg. the 2 in ERCUAl-A2).
A5.8	Filler Metals For Brazing and Braze Welding - Examples BCu-P, RBCuZn-A, BVAg-32 Brazing filler metals are standardized into eight alloy systems: silver, precious metals, aluminum-silicon, copper-phosphorus, copper and copper-zinc, nickel, cobalt, and magnesium filler metals. The primary alloy system is identified according to chemical symbol. 1) At the beginning of the classification a)"R" indicates a brazing filler metal b) "RB" indicates that the filler metal is suitable as a welding rod and as a brazing filler metal c) "BV" indicates a "vacuum grade" filler metals for use in some electronic devices. 2) The letters fillowing trhe "B", "RB" or "BV" are chemical symbols representing the primary alloy composition. CuP in the example refers to a copper-phosphorus alloy. 3) Suffix numerals are used to indicate a particular chemical analysis within an alloy group. 4) A grade suffix is added after any suffix numerals for vacuum grade filler metals as follows: a) Grade 1 indicates the most stringent requirements on the emitter impurities b) Grade 2 indicates less stringent requirements on emitter impurities.
A5.9	Bare Stainless Steel Welding Electrodes and Rods - Example ER309LMo 1) The first two letters of the classification may be: a) ER for solid wires that may be used as electrodes or rods; b) EC for composite cored or stranded wires; or c) EQ for strip electrodes 2) The first three digist of the classification indicate composition. In a few cases, the number of digits may vary but still composition is still indicated. 3) Letters may follow the digits to indicate specific alloy additions. In the case of the example, the letters "L" and "Mo" refer to low carbon grade with a 2.0 to 3.0% molybdenum addition.

AWS	Specification Title, Classification Examples, and Explanation (Continued)
A5.10	Bare Aluminum and Aluminum Alloy Welding Electrodes and Rods - Examples ER4043, R5356, R-C355.0 1) Letters at the beginning of the classification have the following meaning: a) "ER" indicates suitablility as an electrode or a rod b) "R" indicates suitability as welding rod c) A "C" or "A" following the "R" or "ER" is part of the Aluminum Association designation for castings. 2) The four digit number following the leading letters indicate the alloy designation of the Aluminum Association.
A5.11	Nickel and Nickel Alloy Welding Electrodes for Shielded Metal Arc Welding - Example ENiCrMo-3 1. The letter "E" at the beginning of each classification stands for electrode. 2. The chemical symbol "Ni" appears right after the "E" to identify the electrodes as nickel base alloys. 3. Other chemical symbols such as Cr, Cu, Fe, Mo, and Co may follow the "Ni" to group electrodes according to their principal alloying elements. 4. After the chemical symbols, a suffix number is used to identify specific alloys within the same alloy group (eg. ENiMo-1 and ENiMo-3). The numbers are not repeated within the same group.
A5.12	Tungsten And Tungsten Alloy Electrodes For Arc Welding And Cutting - Example EWTh-2 1. The first letter in the designation ,"E", indicates electrode. 2. The following letter "W" indicates that the electrode is primarily tungsten 3. The next letters, "P", "Th", or "Zr" indicate pure tungsten, thoriated tungsten, or zirconiated tungsten, respectively. 4. The numeral at the end of some classifications indicates a different chemical composition or product within the specific group. For example, the "2" in EWTh-2 indicates a 2% thoriated tungsten electrode.
A5.13	Solid Surfacing Welding Rods and Electrodes - Examples RFe5-A, ERCuA1-A2 1. Letters at the beginning of the classification have the following meaning: a) "ER" indicates suitability as an electrode or a rod b) "R" indicates suitability as a welding rod 2. Chemical symbols such as Cr, Cu, Fe, Mo, and Co identify principal alloying elements. 3. Suffix letters and numbers identify specific chemical compositions within the primary alloy system.

AWS	Specification Title, Classification Examples, and Explanation (Continued)
A5.14	Nickel and Nickel Alloy Bare Welding Electrodes and Rods - Examples ERNiCu-7, ERNiCrMo-3.f 1. "ER" at the beginning of each classification indicates that the filler metal may be used as an electrode or a rod. 2. The chemical symbol "Ni" right after the ER identifies the filler metal as a nickel-base alloy. 3. Other symbols such as Cr, Cu, Fe, and Mo group the filler metals according to their principal alloying elements. 4. After the chemical symbols, a suffix number is used to identify separate compositions within the same group (eg. ERNiMo-1 and ERNiMo-3). The numbers are not repeated within the same group.
A5.15	Welding Electrodes and Rods for Cast Iron - Examples ENiFe-Cl-A, ENiCu-B, RCI-A 1. At the beginning of each classification: a) "E" stands for electrode; b) "ER" for a filler metal which is suitable for use as either an electrode or a rod, and c) "R" at the beginning of each classification designation stands for welding rod. 2. The next letters in the filler metal designation are based on the chemical composition of the filler metal or undiluted weld metal. Thus, NiFe is a nickel-iron alloy, NiCu is a nickel-copper alloy, etc. 3. Following the chemical symbols a) a "Cl" to indicate that the filler is intended for cast iron applications b) an "St" to indicate that the filler metal is intended for steel application. c) a "T" is used to indicate a tubular electrode for FCAW and the number after the T indicates whether an external shielding gas is required. (Note that the "CI" and "St" are intended to eliminate confusion with filler metal classifications from other specifications, which are designed for alloys other than cast irons. Two exceptions to this rule are XXXX-A and XXXX-B, where the "A" and "B" indentifiers preceded the introduction of "CI" in the specification.) 4. Where it is necessary to differentiate composition limits of filler metals within the same alloy family , suffix letters such as "A" or "B" are used, as in ENiCu-A and ENiCu-B.

AWS	Specification Title, Classification Examples, and Explanation (Continued)
A5.16	Titanium and Titanium Alloy Welding Rods and Electrodes - Examples ERTi-2, ERTi-6ELI 1. The letter "E" at the beginning of each classification stands for electrode, and the letter "R" stands for welding rod. Since these filler metals are used as electrodes in gas metal arc welding and as rods in gas tungsten arc welding, both letters are used. 2. The chemical symbol "Ti" is used to identify the filler metals as unalloyed titanium or a titanium-base alloy. 3. The numeral following the "Ti" chemical symbol identifies different alloy compositions, and follows the equivalent grade designation of ASTM/ASME specifications for the corresponding base metal. ERTi-15 is an exception to the rule. In the absence of an ASTM/ASME grade number in general usage for Ti-6Al-2Cb1Ta1Mo, the number 15 was arbitrarily assigned to this classification of filler metal. 4. The letters "ELI" at the end of some classifications indicates extra low interstitial content (ie. carbon, oxygen, hydrogen, and nitrogen).

<table>
<tr><th>AWS</th><th>Specification Title, Classification Examples, and Explanation (Continued)</th></tr>
<tr><td>A5.17</td><td>Carbon Steel Electrodes and Fluxes for Submerged Arc Welding -
Examples F6A0-EH14, F7P6-EM12K, and F7P4-ECl

Understanding of the classification method requires separation of the classification into two components:
1. a flux component; and
2. an electrode component

Fluxes are classified before the hyphen, on the basis of the mechanical properties of the weld metal they produce with a given classification of electrode, under the specific test conditions called for the specification.
1. "F" designates a flux.
2. The next single digit represents the minimum tensile strength required of the weld metal in increments of 10000 psi.
3. The third digit is a letter "A" or "P" indicating that the weld metal was tested and classified in the as-welded condition (A) or postweld heat treated condition (P).
4. The digit following the A or P refers to impact testing requirements of weld metal deposited with the flux. A "Z" indicates that no impact requirement is specified. A number indicates that the weld metal satisfies the required 20 ft-lb (27 Joule) Charpy V-notch impact strength at test temperatures given by the digit as follows: 0 = 0°F (-18°C), 2 = -20°F (-29°C), 4 = -40°F (-40°C), 5 = -50°F (-46°C), 6 = -60°F (-51°C), and 8 = -80°F (-62°C).

Electrodes are classified after the hyphen , and represent the filler metal with which the flux will deposit weld metal meeting the specified mechanical properties when tested as called for in the specification.
1. The letter "E" at the beginning of each classification stands for electrode. The letters EC indicate composite electrode.
2. For solid electrodes, the remainder of the designation indicates the chemical composition of the electrode where:
a. "L" indicates that the solid electrode is comparatively low in manganese content.
b. "M" indicates a medium manganese content
c. "H" indicates a comparatively high manganese content.
For composite electrodes, the classification is based upon low dilution weld metal obtained with a particular flux. The numerical suffix following the "EC" refers to composition group.
3. For solid electrodes, the one or two digits after the manganese designator refer to the electrode nominal carbon content.
4. The letter "K", which appears in some designations, indicates that the electrode is made from a heat of silicon-killed steel.</td></tr>
</table>

AWS	Specification Title, Classification Examples, and Explanation (Continued)
A5.18	Carbon Steel Electrodes and Rods Gas Shielded Arc Welding - Example ER70S-2 1. At the beginning of the classification a) "E" designates an electrode b) "ER" indicates that the bare filler metal may be used as an electrode or welding rod. 2. The next two digit number indicates the required minimum tensile strength of the weld metal in multiples of 1000 psi. 3. "S"designates a bare, solid electrode or rod. 4. The suffix number refers to the specific chemical composition shown in the specification.
A5.19	Specification For Magnesium Alloy Welding Electrodes and Rods 1. The prefix R indicates that the material is suitable for use as a welding rod and the prefix E indicates suitability as an electrode. Since some of these filler metals are used as electrodes in gas metal arc welding and as welding rods in oxyfuel or gas tungsten arc welding, both letters may be used. 2. Chemistry is established on the basis of ASTM B 275, Codification of Certain Nonferrous Metals and Alloys, Cast and Wrought.
A5.20	Carbon Steel Electrodes for Flux Cored Arc Welding - Example E70T-6 1. "E" designates an electrode. 2. The first digits, in this case "7", indicates the minimum tensile strength of the deposited metal. For E70T-6, the minimum tensile strength is 70 ksi (70000 psi). 3. The third digit, in this case "0", indicates the primary welding position for which the electrode is designed. a) "0" indicates flat and horizontal positions. b) "1" indicates all positions. 4. "T" stands for tubular, indicating a flux cored electrode. 5. The numerical suffix after the "T" is related to performance and operational characteristics of the electrode, which can be determined from the appendix to AWS A5.20.

<table>
<tr><th>AWS</th><th>Specification Title, Classification Examples, and Explanation (Continued)</th></tr>
<tr><td>A5.21</td><td>Composite Surfacing Welding Rods and Electrodes

To understand the A5.21 classification system, one must first identify two main groups of filler metal:
1. high-speed steels, austenitic manganese steels, and austenitic high chromium irons,
2. tungsten-carbide classifications,

For high-speed steels, austenitic manganese steels, and austenitic high chromium irons, classification examples include RFe5-B and EFeMn-A. The classification system follows:
1. "E" at the beginning of each classification indicates an electrode, "R" indicates a welding rod.
2. Letters immediately after the "E" or "R" are the chemical symbols to group the principal elements in the classification. Thus FeMn is an iron -manganese steel, and FeCr is an iron-chromium alloy, etc.
3. Where more than one clasification is included in a basic group, the individual classifications in the group are identified by the letters A, B, etc., as in EFeMn-A.
4. Further subdividing within an alloy group is done by using a 1, 2, etc., after the last letter.

For tungsten-carbide rods, classification examples include RWC-12/20, RWC-30, EWC20/30 and EWC-40. The classification system is described below.
1. The letters "R" and "E" at the beginning of each classification indicate welding rod and welding electrode, respectively.
2. The "WC" immediately after the "R" or "E" indicates that the filler metal consists of a mild steel tube filled with granules of fused tungsten-carbide.
3. The numbers following the "WC" indicate the mesh size limits for the tungsten-carbide granules. For two numbers separated by a slash, the number preceding the slash indicates the sieve size through which the particles must pass, and the number following the slash indicates the sieve size on which the particles are held. Where only one sieve size is shown, this indicates the size of the screen through which the granules must pass.</td></tr>
<tr><td>A5.22</td><td>Flux Cored Corrosion-Resisting Chromium and Chromium-Nickel Steel Electrodes - Examples E308T-2 and E316LT-2
1. "E" designates an electrode
2. "T" stands for tubular, indicating a flux cored electrode
3. Digits and letters between the "E" and the "T" indicate chemical composition.
4. The suffix after the "T" indicates the shielding medium intended for welding:
a) "1" = carbon dioxide gas plus a flux system.
b) "2" = a mixture of argon with 2 percent oxygen plus a flux system.
c) "3" = self-shielding, with no external shielding gas required.
d) "G" = an electrode with unspecified method of shielding. Properties must be obtained from the manufacturer.</td></tr>
</table>

AWS	Specification Title, Classification Examples, and Explanation (Continued)
A5.23	Low Alloy Steel Electrodes and Fluxes for Submerged Arc Welding - Example F7PO-EL12-A1, F8A4-EA2-A2, F9A10-EA4-A4. Understanding of the classification method requires separation of the classification into three components: 1. a flux component; 2. an electrode component; and 3. a weld metal chemistry component. Fluxes are classified before the first hyphen, on the basis of the mechanical properties of the weld metal they produce with a given classification of electrode, under the specific test conditions called for in the specification. 1. "F" designates a flux. 2. The one or two digits after "F" represent the minimum weld metal tensile strength in 10000 psi increments. 3. The third element of the flux classification is a letter "A" or "P" indicating that the weld metal was tested and classified in the as-welded condition (A) or postweld heat treated condition (P). 4. The digit that follows the A or P will be either a "Z" indicating that no impact test requirement is specified, or a digit indicating that the weld metal satisfies the required 20 ft-Ib (27 Joule) Charpy V-notch impact strength at the test temperature indicated by the digit where: 0 = 0°F (-18°C), 2 = -20°F (-29°C), 4 = -40°F (-40°C), 5 = -50°F (-46°C), 6 = -60°F (-51°C), 8 = -80°F (-62°C). The central part of the classification (eg. EL12, ENi3, ECB3. or ECM10, refers to the electrode classification with which the flux will produce weld metal that meets the specified mechanical properties when tested as called for in the specification. 1. "E" at the beginning of each classification stands for electrode. The letters."EC" indicate composite electrode. 2. The remaining letters and numbers indicate the chemical composition of the electrode, or, in the case of composite electrodes, of the undiluted weld metal obtained with a particular flux. a. For the EL12 and EM12K classifications, chemical composition requirements are the same as AWS A5.17. b. For other electrodes, compositions are shown in Table 1 of the specification. c. "N" indicates the electrode is intended for nuclear applications. d. "G" indicates the filler metal is of a "general" classification. It allows space for a useful filler metal, which otherwise would have to wait for revision of the specification to be classified. Consequently, two filler metals bearing the "G" classification, may be quite different in some respect such as chemical composition. The final component of the classification (eg. A1, A2, A4) refers to the chemical composition requirements for undiluted weld metal, which may be obtained from Table 2 of the specification.

AWS	Specification Title, Classification Examples, and Explanation (Continued)
A5.24	Zirconium and Zirconium Alloy Welding Electrodes and Rods - Example ERZr3 1. "ER" at the beginning of each designation indicates that the filler metal may be used as a welding electrode or rod. 2. "Zr" indicates that the filler metals have a zirconium base. 3. The characters following the "Zr" chemical symbol identify the nominal composition of the filler metal (see specification).
A5.25	Carbon And Low Alloy Steel Electrodes And Fluxes For Electroslag Welding - Examples FES60-EH14-EW, FES72-EWT2 Understanding of the classification method requires separation of the classification into two components: 1. a flux component; and 2. an electrode component. Fluxes are classified before the first hyphen (eg. FES60, FES72), on the basis of the mechanical properties of the weld metal they produce with a given classification of electrode, under the specific test conditions called for in the specification. 1. "FES" designates a flux for electroslag welding. 2. The digit after "FES" represents the minimum weld metal tensile strength in 10000 psi increments. 3. The digit after the strength designator will be either a "Z" indicating that no impact test requirement is specified, or a digit indicating that the weld metal satisfies the required 15 ft-lb (20 Joule) Charpy V-notch impact strength at the test temperature indicated by the digit where: 0 = 0°F (-18°C), 2 = -20°F (-29°C), 4 = -40°F (-40°C), 5 = -50°F (-46°C), 6 = -60°F (-51°C), 8 = -80°F (-62°C). Electrodes are classified after the first hypen (eg. EH14-EW, EWT2), on the basis of chemical composition. 1. "E" at the beginning of each classification stands for electrode. 2. The remaining letters and numbers indicate the chemical composition of the electrode, or, in the case of composite electrodes, of the undiluted weld metal obtained with a particular flux. a) "M" indicates medium manganese content. b) "H" indicates high manganese content. c) Digits following the M or H indicate the nominal carbon content of the electrode. d) "K", when present, indicates that the electrode is made from a silicon killed steel. e) "EW" indicates a solid wire electrode. c) "WT" indicates a composite electrode. d) "G" indicates the filler metal is of a "general" classification. It allows space for a useful filler metal, which otherwise would have to wait for revision of the specification to be classified. Consequently, two filler metals bearing the "G" classification, may be quite different in some respect such as chemical composition.

AWS	Specification Title, Classification Examples, and Explanation (Continued)
A5.26	Carbon And Low Alloy Steel Electrodes For Electrogas Welding - Example EG62S-1 1. "EG" at the beginning of each classification indicates that the electrode is intented for electogas welding. 2. The first digit following the "EG" represents the minimum tensile strength of the weld metal in units of 10000 psi. 3. The next letter, either an "S" or a "T", indicates that the electrode is solid (S) or composite flux cored or metal cored (T). 4. The designators after the hyphen refer to chemical composition of the electrode, or, in the case of composite electrodes, of the undiluted weld metal obtained with a particular flux, and the type or absence of shielding gas. A "G" indicates the filler metal is of a "general" classification. It allows space for a useful filler metal, which otherwise would have to wait for revision of the specification to be classified. Consequently, two filler metals bearing the "G" classification, may be quite different in some respect such as chemical composition.
A5.27	Copper and Copper Alloy Gas Welding Rods - Examples RCuZn-C, ERCu, and RBCuZn-D 1. "R", "ER", and "RB" at the beginning of each classification indicates that the welding consumable may be an oxyfuel gas welding rod, either an electrode or brazing filler metal, and either a welding rod or a brazing filler metal. 2. "Cu" is used to identify the welding rods as copper-base alloys and the additional chemical symbols indicates the principal alloying element of each group. 3. Where more than one classification is included in a basic group, the individual classifications in the group are identified by the letters "A", "B", "C," etc.
A5.28	Low Alloy Steel Filler Metals - Examples ER80S-B2 and E80C-B2. 1. "E"designates an electrode, as in other specifications. "ER" at the beginning of a classification indicates that the filler metal may be used as an electrode or a welding rod. 2. The number 80 indicates the required minimum tensile strength of weld metal in multiples of 1000 psi. Three digits are used for weld metal of 100 000 psi tensile strength and higher. 3. "S" designates a bare solid electrode or rod, while "C" designates a composite metal cored or stranded electrode. 4. The suffix B2 indicates a particular classification based on as-manufactured chemical composition.

AWS	Specification Title, Classification Examples, and Explanation (Continued)
A5.29	Low Alloy Electrodes for Flux Cored Arc Welding - Examples E80T5-B2L, E100T5-D2, E120T5-K4 1. "E" stands for electrode. 2. The digit or digits between the "E" and the first digit before the "T" indicates the minimum tensile strength of the deposited metal in increments of 10000 pst. 3. The digit immediately before the "T" indicates the primary welding position for which the electrode is designed. a) "0" indicates flat and horizontal positions. b) "1" indicates all positions. 4. "T" stands for tubular, indicating a flux cored electrode. 5. The numerical suffix after the "T" is related to performance and operational characteristics of the electrode, which can be determined from the appendix to AWS A5.29.
A5.30	Consumable Inserts - Example IN308 1. The prefix "IN" designates a consumable insert. 2. The numbers 308 designate the chemical composition. Note that, while the solid products are classified on the basis of chemical composition, their cross-sectional configurations are another consideration that must be selected and specified when ordering.
A5.31	Fluxes For Brazing And Braze Welding a) "FB" indicates Flux for Brazing or Braze Welding b) The third character refers to group of applicable base metals as listed in the standard. c) A fourth character represents a change in form and attendant composition, within the broader base metal classification.

Appendix 2

Engineering Data

ASME Piping Codes

B31G - Manual for Determining the Remaining Strength of Corroded Pipelines
B31.1 - Power Piping
B31.2 - Fuel Gas Piping
B31.3 - Process Piping
B31.4 - Liquid Transportation Systems for Hydrocarbons, Liquid Petroleum Gas, Anhydrous Ammonia, and Alcohols
B31.5 - Refrigeration Piping
B31.8 - Gas Transportation and Distribution Piping Systems
B31.9 - Building Services Piping
B31.11 - Slurry Transportation Piping Systems

ASME Boiler and Pressure Vessel Code

Section I - Rules for Construction of Power Boilers
Section II - Materials
- Part A - Ferrous Materials Specifications
- Part B - Nonferrous Materials Specifications
- Part C - Specifications for Welding Rods, Electrodes, and Filler Metals
- Part D - Properties

Section III - Subsection NCA - General Requirements for Division 1 and Division 2
Section III - Division 1
- Subsection NB - Class 1 Components
- Subsection NC - Class 2 Components
- Subsection ND - Class 3 Components
- Subsection NE - Class MC Components
- Subsection NF - Supports
- Subsection NG - Core Support Structures
- Subsection NH - Class 1 Components in Elevated Temperature Service
- Appendices

Section III - Division 2 - Code for Concrete Reactor Vessels and Containments
Section IV - Rules for Construction of Heating Boilers
Section V - Nondestructive Examination
Section VI - Recommended Rules for the Care and Operation of Heating Boilers
Section VII - II - Recommended Guidelines for the Care of Power Boilers
Section VIII - Rules for Construction of Pressure Vessels
- Division 1
- Division 2 - Alternative Rules

Section IX - Welding and Brazing Qualifications
Section X - Fiber-Reinforced Plastic Pressure Vessels
Section XI - Rules for Inservice Inspection of Nuclear Power Plant Components

DIMENSIONS OF WELDED AND SEAMLESS PIPE								
Nominal Pipe Size, in.	Outside Diameter	Nominal Wall Thickness (in) For						
		Schedule 5S	Schedule 10S	Schedule 10	Schedule 20	Schedule 30	Schedule Standard	Schedule 40
⅛	0.405	---	0.049	---	---	---	0.068	0.068
¼	0.540	---	0.065	---	---	---	0.088	0.088
⅜	0.675	---	0.065	---	---	---	0.091	0.091
½	0.840	0.065	0.083	---	---	---	0.109	0.109
¾	1.050	0.065	0.083	---	---	---	0.113	0.113
1	1.315	0.065	0.109	---	---	---	0.133	0.133
1¼	1.660	0.065	0.109	---	---	---	0.140	0.140
1½	1.900	0.065	0.109	---	---	---	0.145	0.145
2	2.375	0.065	0.109	---	---	---	0.154	0.154
2½	2.875	0.083	0.120	---	---	---	0.203	0.203
3	3.5	0.083	0.120	---	---	---	0.216	0.216
3½	4.0	0.083	0.120	---	---	---	0.226	0.226
4	4.5	0.083	0.120	---	---	---	0.237	0.237
5	5.563	0.109	0.134	---	---	---	0.258	0.258
6	6.625	0.109	0.134	---	---	---	0.280	0.280
8	8.625	0.109	0.148	---	0.250	0.277	0.322	0.322
10	10.75	0.134	0.165	---	0.250	0.307	0.365	0.365
12	12.75	0.156	0.180	---	0.250	0.330	0.375	0.406
14 OD	14.0	0.156	0.188	0.250	0.312	0.375	0.375	0.438
16 OD	16.0	0.165	0.188	0.250	0.312	0.375	0.375	0.500
18 OD	18.0	0.165	0.188	0.250	0.312	0.438	0.375	0.562
20 OD	20.0	0.188	0.218	0.250	0.375	0.500	0.375	0.594
22 OD	22.0	0.188	0.218	0.250	0.375	0.500	0.375	---
24 OD	24.0	0.218	0.250	0.250	0.375	0.562	0.375	0.688
26 OD	26.0	---	---	0.312	0.500	---	0.375	---
28 OD	28.0	---	---	0.312	0.500	0.625	0.375	---
30 OD	30.0	0.250	0.312	0.312	0.500	0.625	0.375	---
32 OD	32.0	---	---	0.312	0.500	0.625	0.375	0.688
34 OD	34.0	---	---	0.312	0.500	0.625	0.375	0.688
36 OD	36.0	---	---	0.312	0.500	0.625	0.375	0.750
42 OD	42.0	---	---	---	---	---	0.375	---

See next table for heavier wall thicknesses; all units are inches.

DIMENSIONS OF WELDED AND SEAMLESS PIPE									
Nominal Pipe Size, in.	**Outside Diameter**	**Nominal Wall Thickness (in) For**							
		Schedule 60	**Extra Strong**	**Schedule 80**	**Schedule 100**	**Schedule 120**	**Schedule 140**	**Schedule 160**	**XX Strong**
1/8	0.405	---	0.095	0.095	---	---	---	---	---
1/4	0.540	---	0.119	0.119	---	---	---	---	---
3/8	0.675	---	0.126	0.126	---	---	---	---	---
1/2	0.840	---	0.147	0.147	---	---	---	0.188	0.294
3/4	1.050	---	0.154	0.154	---	---	---	0.219	0.308
1	1.315	---	0.179	0.179	---	---	---	0.250	0.358
1 1/4	1.660	---	0.191	0.191	---	---	---	0.250	0.382
1 1/2	1.900	---	0.200	0.200	---	---	---	0.281	0.400
2	2.375	---	0.218	0.218	---	---	---	0.344	0.436
2 1/2	2.875	---	0.276	0.276	---	---	---	0.375	0.552
3	3.5	---	0.300	0.300	---	---	---	0.438	0.600
3 1/2	4.0	---	0.318	0.318	---	---	---	---	---
4	4.5	---	0.337	0.337	---	0.438	---	0.531	0.674
5	5.563	---	0.375	0.375	---	0.500	---	0.625	0.750
6	6.625	---	0.432	0.432	---	0.562	---	0.719	0.864
8	8.625	0.406	0.500	0.500	0.594	0.719	0.812	0.906	0.875
10	10.75	0.500	0.500	0.594	0.719	0.844	1.000	1.125	1.000
12	12.75	0.562	0.500	0.688	0.844	1.000	1.125	1.312	1.000
14 OD	14.0	0.594	0.500	0.750	0.938	1.094	1.250	1.406	---
16 OD	16.0	0.656	0.500	0.844	1.031	1.219	1.438	1.594	---
18 OD	18.0	0.750	0.500	0.938	1.156	1.375	1.562	1.781	---
20 OD	20.0	0.812	0.500	1.031	1.281	1.500	1.750	1.969	---
22 OD	22.0	0.875	0.500	1.125	1.375	1.625	1.875	2.125	---
24 OD	24.0	0.969	0.500	1.218	1.531	1.812	2.062	2.344	---
26 OD	26.0	---	0.500	---	---	---	---	---	---
28 OD	28.0	---	0.500	---	---	---	---	---	---
30 OD	30.0	---	0.500	---	---	---	---	---	---
32 OD	32.0	---	0.500	---	---	---	---	---	---
34 OD	34.0	---	0.500	---	---	---	---	---	---
36 OD	36.0	---	0.500	---	---	---	---	---	---
42 OD	42.0	---	0.500	---	---	---	---	---	---

All units are inches.

To Convert From	To	Multiply By	To Convert From	To	Multiply By
Angle			**Mass per unit length**		
degree	rad	1.745 329 E -02	lb/ft	kg/m	1.488 164 E + 00
Area			lb/in.	kg/m	1.785 797 E + 01
in.^{2}	mm^2	6.451 600 E + 02	**Mass per unit time**		
in.^{2}	cm^2	6.451 600 E + 00	lb/h	kg/s	1.259 979 E - 04
in.^{2}	m^2	6.451 600 E - 04	lb/min	kg/s	7.559 873 E - 03
ft^2	m^2	9.290 304 E - 02	lb/s	kg/s	4.535 924 E - 01
Bending moment or torque			**Mass per unit volume (includes density)**		
lbf - in.	N - m	1.129 848 E - 01	g/cm^3	kg/m^3	1.000 000 E + 03
lbf - ft	N - m	1.355 818 E + 00	lb/ft^3	g/cm^3	1.601 846 E - 02
kgf - m	N - m	9.806 650 E + 00	lb/ft^3	kg/m^3	1.601 846 E + 01
ozf - in.	N-m	7.061 552 E - 03	lb/in.^{3}	g/cm^3	2.767 990 E + 01
Bending moment or torque per unit length			lb/in.^{3}	kg/m^3	2.767 990 E + 04
lbf - in./in.	N - m/m	4.448 222 E + 00	**Power**		
lbf - ft/in.	N - m/m	5.337 866 E + 01	Btu/s	kW	1.055 056 E + 00
Corrosion rate			Btu/min	kW	1.758 426 E - 02
mils/yr	mm/yr	2.540 000 E - 02	Btu/h	W	2.928 751 E - 01
mils/yr	μ/yr	2.540 000 E + 01	erg/s	W	1.000 000 E - 07
Current density			ft - lbf/s	W	1.355 818 E + 00
A/in.^{2}	A/cm^2	1.550 003 E - 01	ft - lbf/min	W	2.259 697 E - 02
A/in.^{2}	A/mm^2	1.550 003 E - 03	ft - lbf/h	W	3.766 161 E - 04
A/ft^2	A/m^2	1.076 400 E + 01	hp (550 ft - lbf/s)	kW	7.456 999 E - 01
Electricity and magnetism			hp (electric)	kW	7.460 000 E - 01
gauss	T	1.000 000 E - 04	W/in.^{2}	W/m^2	1.550 003 E + 03
maxwell	μWb	1.000 000 E - 02	**Pressure (fluid)**		
mho	S	1.000 000 E + 00	atm (standard)	Pa	1.013 250 E + 05
Oersted	A/m	7.957 700 E + 01	bar	Pa	1.000 000 E + 05
Ω - cm	Ω - m	1.000 000 E - 02	in. Hg (32 F)	Pa	3.386 380 E + 03
Ω circular - mil/ft	μΩ - m	1.662 426 E - 03	in. Hg (60 F)	Pa	3.376 850 E + 03
Energy (impact other)			lbf/in.^{2} (psi)	Pa	6.894 757 E + 03
ft - lbf	J	1.355 818 E + 00	torr (mm Hg, 0 C)	Pa	1.333 220 E + 02
Btu (thermochemical)	J	1.054 350 E + 03	**Specific heat**		
cal (thermochemical)	J	4.184 000 E + 00	Btu/lb - F	J/kg - K	4.186 800 E + 03
kW - h	J	3.600 000 E + 06	cal/g - C	J/kg - K	4.186 800 E + 03
W - h	J	3.600 000 E + 03	**Stress (force per unit area)**		
Flow rate			tonf/in.^{2} (tsi)	MPa	1.378 951 E + 01
ft^3/h	L/min	4.719 475 E - 01	kgf/mm^2	MPa	9.806 650 E + 00
ft^3/min	L/min	2.831 000 E + 01	ksi	MPa	6.894 757 E + 00
gal/h	L/min	6.309 020 E - 02	lbf/in.^{2} (psi)	MPa	6.894 757 E - 03
gal/min	L/min	3.785 412 E + 00	MN/m^2	MPa	1.000 000 E + 00
Force			**Temperature**		
lbf	N	4.448 222 E + 00	F	C	5/9 (F - 32)
kip (1000 lbf)	N	4.448 222 E + 03	R	K	5/9
tonf	kN	8.896 443 E + 00	**Thermal conductivity**		
kgf	N	9.806 650 E + 00	Btu - in./s - ft^2 - F	W/m - K	5.192 204 E + 02
			Btu/ft - h - F	W/m - K	1.730 735 E + 00

To Convert From	To	Multiply By	To Convert From	To	Multiply By
Force per unit length			**Thermal conductivity (Con't)**		
lbf/ft	N/m	1.459 390 E + 01	Btu - in./h . ft^2 - F	W/m - K	1.442 279 E - 01
lbf/in.	N/m	1.751 268 E + 02	cal/cm - s - C	W/m - K	4.184 000 E + 02
Fracture toughness			**Thermal expansion**		
ksi √in.	MPa √m	1.098 800 E + 00	in./in. - C	m/m - K	1.000 000 E + 00
Heat content			in./in. - F	m/m - K	1.800 000 E + 00
Btu/lb	kJ/kg	2.326 000 E + 00	**Velocity**		
cal/g	kJ/kg	4.186 800 E + 00	ft/h	m/s	8.466 667 E - 05
Heat input			ft/min	m/s	5.080 000 E - 03
J/in.	J/m	3.937 008 E + 01	ft/s	m/s	3.048 000 E - 01
kJ/in.	kJ/m	3.937 008 E + 01	in./s	m/s	2.540 000 E - 02
Length			km/h	m/s	2.777 778 E - 01
A	nm	1.000 000 E - 01	mph	km/h	1.609 344 E + 00
μin.	μm	2.540 000 E - 02	**Velocity of rotation**		
mil	μm	2.540 000 E + 01	rev/min (rpm)	rad/s	1.047 164 E - 01
in.	mm	2.540 000 E + 01	rev/s	rad/s	6.283 185 E + 00
in.	cm	2.540 000 E + 00	**Viscosity**		
ft	m	3.048 000 E - 01	poise	Pa - s	1.000 000 E - 01
yd	m	9.144 000 E -01	stokes	m^2/S	1.000 000 E - 04
mile	km	1.609 300 E + 00	ft^2/s	m^2/s	9.290 304 E - 02
Mass			$in.^2$/s	mm^2/s	6.451 600 E + 02
oz	kg	2.834 952 E - 02	**Volume**		
lb	kg	4.535 924 E - 01	$in.^3$	m^3	1.638 706 E - 05
ton (short 2000 lb)	kg	9.071 847 E + 02	ft^3	m^3	2.831 685 E - 02
ton (short 2000 lb)	kg x 10^3	9.071 847 E - 01	fluid oz	m^3	2.957 353 E - 05
ton (long 2240 lb)	kg	1.016 047 E + 03	gal (U.S. liquid)	m^3	3.785 412 E - 03
kg x 10^3 = 1 metric ton			**Volume per unit time**		
Mass per unit area			ft^3/min	m^3/S	4.719 474 E - 04
oz/$in.^2$	kg/m^2	4.395 000 E + 01	ft^3/S	m^3/s	2.831 685 E - 02
oz/ft^2	kg/m^2	3.051 517 E - 01	$in.^3$/min	m^3/S	2.731 177 E - 07
oz/yd^2	kg/m^2	3.390 575 E - 02	**Wavelength**		
lb/ft^2	kg/m^2	4.882 428 E + 00	A	nm	1.000 000 E - 01

SI PREFIXES

Prefix	Symbol	Exponential Expression	Multiplication Factor
peta	P	10^{15}	1 000 000 000 000 000
tera	T	10^{12}	1 000 000 000 000
giga	G	10^{9}	1 000 000 000
mega	M	10^{6}	1 000 000
kilo	k	10^{3}	1 000
hecto	h	10^{2}	100
deka	da	10^{1}	10
Base Unit	---	10^{0}	1
deci	d	10^{-1}	0.1
centi	c	10^{-2}	0.01
milli	m	10^{-3}	0.001
micro	μ	10^{-6}	0.000 001
nano	n	10^{-9}	0.000 000 001
pico	p	10^{-12}	0.000 000 000 001
femto	f	10^{-15}	0.000 000 000 000 001

APPROXIMATE HARDNESS CONVERSION NUMBERS FOR NONAUSTENITIC STEELS[a, b]								
Rockwell C 150 kgf Diamond HRC	Vickers HV	Brinell 3000 kgf 10mm ball[c] HB	Knoop 500 gf HK	Rockwell A 60 kgf Diamond HRA	Rockwell Superficial Hardness 15 kgf Diamond HR15N	30 kgf Diamond HR30N	45 kgf Diamond HR45N	Approx. Tensile Strength ksi (MPa)
68	940	---	920	85.6	93.2	84.4	75.4	---
67	900	---	895	85.0	92.9	83.6	74.2	---
66	865	---	870	84.5	92.5	82.8	73.3	---
65	832	739[d]	846	83.9	92.2	81.9	72.0	---
64	800	722[d]	822	83.4	91.8	81.1	71.0	---
63	772	706[d]	799	82.8	91.4	80.1	69.9	---
62	746	688[d]	776	82.3	91.1	79.3	68.8	---
61	720	670[d]	754	81.8	90.7	78.4	67.7	---
60	697	654[d]	732	81.2	90.2	77.5	66.6	---
59	674	634[d]	710	80.7	89.8	76.6	65.5	351 (2420)
58	653	615	690	80.1	89.3	75.7	64.3	338 (2330)
57	633	595	670	79.6	88.9	74.8	63.2	325 (2240)
56	613	577	650	79.0	88.3	73.9	62.0	313 (2160)
55	595	560	630	78.5	87.9	73.0	60.9	301 (2070)
54	577	543	612	78.0	87.4	72.0	59.8	292 (2010)
53	560	525	594	77.4	86.9	71.2	58.6	283 (1950)
52	544	512	576	76.8	86.4	70.2	57.4	273 (1880)
51	528	496	558	76.3	85.9	69.4	56.1	264 (1820)
50	513	482	542	75.9	85.5	68.5	55.0	255 (1760)
49	498	468	526	75.2	85.0	67.6	53.8	246 (1700)
48	484	455	510	74.7	84.5	66.7	52.5	238 (1640)
47	471	442	495	74.1	83.9	65.8	51.4	229 (1580)
46	458	432	480	73.6	83.5	64.8	50.3	221 (1520)
45	446	421	466	73.1	83.0	64.0	49.0	215 (1480)
44	434	409	452	72.5	82.5	63.1	47.8	208 (1430)
43	423	400	438	72.0	82.0	62.2	46.7	201 (1390)
42	412	390	426	71.5	81.5	61.3	45.5	194 (1340)
41	402	381	414	70.9	80.9	60.4	44.3	188 (1300)
40	392	371	402	70.4	80.4	59.5	43.1	182 (1250)
39	382	362	391	69.9	79.9	58.6	41.9	177 (1220)
38	372	353	380	69.4	79.4	57.7	40.8	171 (1180)
37	363	344	370	68.9	78.8	56.8	39.6	166 (1140)
36	354	336	360	68.4	78.3	55.9	38.4	161 (1110)
35	345	327	351	67.9	77.7	55.0	37.2	156 (1080)
34	336	319	342	67.4	77.2	54.2	36.1	152 (1050)
33	327	311	334	66.8	76.6	53.3	34.9	149 (1030)
32	318	301	326	66.3	76.1	52.1	33.7	146 (1010)
31	310	294	318	65.8	75.6	51.3	32.5	141 (970)
30	302	286	311	65.3	75.0	50.4	31.3	138 (950)
29	294	279	304	64.6	74.5	49.5	30.1	135 (930)
28	286	271	297	64.3	73.9	48.6	28.9	131 (900)
27	279	264	290	63.8	73.3	47.7	27.8	128 (880)
26	272	258	284	63.3	72.8	46.8	26.7	125 (860)
25	266	253	278	62.8	72.2	45.9	25.5	123 (850)

APPROXIMATE HARDNESS CONVERSION NUMBERS FOR NONAUSTENITIC STEELS[a, b] (Continued)

Rockwell C 150 kgf Diamond HRC	Vickers HV	Brinell 3000 kgf 10mm ball[c] HB	Knoop 500 gf HK	Rockwell A 60 kgf Diamond HRA	Rockwell Superficial Hardness 15 kgf Diamond HR15N	Rockwell Superficial Hardness 30 kgf Diamond HR30N	Rockwell Superficial Hardness 45 kgf Diamond HR45N	Approx. Tensile Strength ksi (MPa)
24	260	247	272	62.4	71.6	45.0	24.3	119 (820)
23	254	243	266	62.0	71.0	44.0	23.1	117 (810)
22	248	237	261	61.5	70.5	43.2	22.0	115 (790)
21	243	231	256	61.0	69.9	42.3	20.7	112 (770)
20	238	226	251	60.5	69.4	41.5	19.6	110 (760)

a. This table gives the approximate interrelationships of hardness values and approximate tensile strength of steels. It is possible that steels of various compositions and processing histories will deviate in hardness-tensile strength relationship from the data presented in this table. The data in this table should not be used for austenitic stainless steels, but have been shown to be applicable for ferritic and martensitic stainless steels. Where more precise conversions are required, they should be developed specially for each steel composition, heat treatment, and part.

b. All relative hardness values in this table are averages of tests on various metals whose different properties prevent establishment of exact mathematical conversions. These values are consistent with ASTM A 370-91 for nonaustenitic steels. It is recommended that ASTM standards A 370, E 140, E 10, E 18, E 92, E 110 and E 384, involving hardness tests on metals, be reviewed prior to interpreting hardness conversion values.

c. Carbide ball, 10mm.

d. This Brinell hardness value is outside the recommended range for hardness testing in accordance with ASTM E 10.

APPROXIMATE HARDNESS CONVERSION NUMBERS FOR NONAUSTENITIC STEELS[a, b]

Rockwell B 100 kgf 1/16" ball HRB	Vickers HV	Brinell 3000 kgf 10 mm HB	Knoop 500 gf HK	Rockwell A 60 kgf Diamond HRA	Rockwell Superficial Hardness 15 kgf 1/16" ball HR15T	Rockwell Superficial Hardness 30 kgf 1/16" ball HR30T	Rockwell Superficial Hardness 45 kgf 1/16" ball HR45T	Approx. Tensile Strength ksi (MPa)
100	240	240	251	61.5	93.1	83.1	72.9	116 (800)
99	234	234	246	60.9	92.8	82.5	71.9	114 (785)
98	228	228	241	60.2	92.5	81.8	70.9	109 (750)
97	222	222	236	59.5	92.1	81.1	69.9	104 (715)
96	216	216	231	58.9	91.8	80.4	68.9	102 (705)
95	210	210	226	58.3	91.5	79.8	67.9	100 (690)
94	205	205	221	57.6	91.2	79.1	66.9	98 (675)
93	200	200	216	57.0	90.8	78.4	65.9	94 (650)
92	195	195	211	56.4	90.5	77.8	64.8	92 (635)
91	190	190	206	55.8	90.2	77.1	63.8	90 (620)
90	185	185	201	55.2	89.9	76.4	62.8	89 (615)
89	180	180	196	54.6	89.5	75.8	61.8	88 (605)
88	176	176	192	54.0	89.2	75.1	60.8	86 (590)
87	172	172	188	53.4	88.9	74.4	59.8	84 (580)
86	169	169	184	52.8	88.6	73.8	58.8	83 (570)
85	165	165	180	52.3	88.2	73.1	57.8	82 (565)
84	162	162	176	51.7	87.9	72.4	56.8	81 (560)
83	159	159	173	51.1	87.6	71.8	55.8	80 (550)
82	156	156	170	50.6	87.3	71.1	54.8	77 (530)
81	153	153	167	50.0	86.9	70.4	53.8	73 (505)
80	150	150	164	49.5	86.6	69.7	52.8	72 (495)
79	147	147	161	48.9	86.3	69.1	51.8	70 (485)
78	144	144	158	48.4	86.0	68.4	50.8	69 (475)
77	141	141	155	47.9	85.6	67.7	49.8	68 (470)
76	139	139	152	47.3	85.3	67.1	48.8	67 (460)
75	137	137	150	46.8	85.0	66.4	47.8	66 (455)

APPROXIMATE HARDNESS CONVERSION NUMBERS FOR NONAUSTENITIC STEELS[a, b] (Continued)								
Rockwell B 100 kgf 1/16" ball HRB	Vickers HV	Brinell 3000 kgf 10 mm HB	Knoop 500 gf HK	Rockwell A 60 kgf Diamond HRA	Rockwell Superficial Hardness			Approx. Tensile Strength ksi (MPa)
					15 kgf 1/16" ball HR15T	30 kgf 1/16" ball HR30T	45 kgf 1/16" ball HR45T	
74	135	135	147	46.3	84.7	65.7	46.8	65 (450)
73	132	132	145	45.8	84.3	65.1	45.8	64 (440)
72	130	130	143	45.3	84.0	64.4	44.8	63 (435)
71	127	127	141	44.8	83.7	63.7	43.8	62 (425)
70	125	125	139	44.3	83.4	63.1	42.8	61 (420)
69	123	123	137	43.8	83.0	62.4	41.8	60 (415)
68	121	121	135	43.3	82.7	61.7	40.8	59 (405)
67	119	119	133	42.8	82.4	61.0	39.8	58 (400)
66	117	117	131	42.3	82.1	60.4	38.7	57 (395)
65	116	116	129	41.8	81.8	59.7	37.7	56 (385)
64	114	114	127	41.4	81.4	59.0	36.7	---
63	112	112	125	40.9	81.1	58.4	35.7	---
62	110	110	124	40.4	80.8	57.7	34.7	---
61	108	108	122	40.0	80.5	57.0	33.7	---
60	107	107	120	39.5	80.1	56.4	32.7	---
59	106	106	118	39.0	79.8	55.7	31.7	---
58	104	104	117	38.6	79.5	55.0	30.7	---
57	103	103	115	38.1	79.2	54.4	29.7	---
56	101	101	114	37.7	78.8	53.7	28.7	---
55	100	100	112	37.2	78.5	53.0	27.7	---
54	---	---	111	36.8	78.2	52.4	26.7	---
53	---	---	110	36.3	77.9	51.7	25.7	---
52	---	---	109	35.9	77.5	51.0	24.7	---
51	---	---	108	35.5	77.2	50.3	23.7	---
50	---	---	107	35.0	76.9	49.7	22.7	---
49	---	---	106	34.6	76.6	49.0	21.7	---
48	---	---	105	34.1	76.2	48.3	20.7	---
47	---	---	104	33.7	75.9	47.7	19.7	---
46	---	---	103	33.3	75.6	47.0	18.7	---
45	---	---	102	32.9	75.3	46.3	17.7	---
44	---	---	101	32.4	74.9	45.7	16.7	---
43	---	---	100	32.0	74.6	45.0	15.7	---
42	---	---	99	31.6	74.3	44.3	14.7	---
41	---	---	98	31.2	74.0	43.7	13.6	---
40	---	---	97	30.7	73.6	43.0	12.6	---
39	---	---	96	30.3	73.3	42.3	11.6	---
38	---	---	95	29.9	73.0	41.6	10.6	---
37	---	94	29.5	78.0	41.0	9.6		---
36	---	---	93	29.1	72.3	40.3	8.6	---
35	---	---	92	28.7	72.0	39.6	7.6	---
34	---	---	91	28.2	71.7	39.0	6.6	---
33	---	---	90	27.8	71.4	38.3	5.6	---
32	---	---	89	27.4	71.0	37.6	4.6	---
31	---	---	88	27.0	70.7	37.0	3.6	---
30	---	---	87	26.6	70.4	36.3	2.6	---

APPROXIMATE HARDNESS CONVERSION NUMBERS FOR NONAUSTENITIC STEELS[a, b] (Continued)

a. This table gives the approximate interrelationships of hardness values and approximate tensile strength of steels. It is possible that steels of various compositions and processing histories will deviate in hardness-tensile strength relationship from the data presented in this table. The data in this table should not be used for austenitic stainless steels, but have been shown to be applicable for ferritic and martensitic stainless steels. Where more precise conversions are required, they should be developed specially for each steel composition, heat treatment, and part.

b. All relative hardness values in this table are averages of tests on various metals whose different properties prevent establishment of exact mathematical conversions. These values are consistent with ASTM A 370-91 for nonaustenitic steels. It is recommended that ASTM standards A 370, E 140, E 10, E 18, E 92, E 110 and E 384, involving hardness tests on metals, be reviewed prior to interpreting hardness conversion values.

APPROXIMATE HARDNESS NUMBERS FOR AUSTENITIC STEELS[a]

Rockwell C 150 kgf, Diamond HRC	Rockwell A 60 kgf, Diamond HRA	Rockwell Superficial Hardness		
		15 kgf, Diamond HR15N	30 kgf, Diamond HR30N	45 kgf, Diamond HR45N
48	74.4	84.1	66.2	52.1
47	73.9	83.6	65.3	50.9
46	73.4	83.1	64.5	49.8
45	72.9	82.6	63.6	48.7
44	72.4	82.1	62.7	47.5
43	71.9	81.6	61.8	46.4
42	71.4	81.0	61.0	45.2
41	70.9	80.5	60.1	44.1
40	70.4	80.0	59.2	43.0
39	69.9	79.5	58.4	41.8
38	69.3	79.0	57.5	40.7
37	68.8	78.5	56.6	39.6
36	68.3	78.0	55.7	38.4
35	67.8	77.5	54.9	37.3
34	67.3	77.0	54.0	36.1
33	66.8	76.5	53.1	35.0
32	66.3	75.9	52.3	33.9
31	65.8	75.4	51.4	32.7
30	65.3	74.9	50.5	31.6
29	64.8	74.4	49.6	30.4
28	64.3	73.9	48.8	29.3
27	63.8	73.4	47.9	28.2
26	63.3	72.9	47.0	27.0
25	62.8	72.4	46.2	25.9
24	62.3	71.9	45.3	24.8
23	61.8	71.3	44.4	23.6
22	61.3	70.8	43.5	22.5
21	60.8	70.3	42.7	21.3
20	60.3	69.8	41.8	20.2

a. All relative hardness values in this table are averages of tests on various metals whose different properties prevent establishment of exact mathematical conversions. These values are consistent with ASTM A 370-91 for austenitic steels. It is recommended that ASTM standards A 370, E 140, E 10, E 18, E 92, E 110 and E 384, involving hardness tests on metals, be reviewed prior to interpreting hardness conversion values.

APPROXIMATE HARDNESS CONVERSION VALUES FOR AUSTENITIC STEELS[a]

Rockwell B 100 kgf 1/16" ball HRB	Brinell Indentation Diameter, mm	Brinell 3000 kgf 10 mm Ball HB	Rockwell A 60 kgf Diamond HRA	Rockwell Superficial Hardness 15 kgf 1/16" ball HR15T	Rockwell Superficial Hardness 30 kgf 1/16" ball HR30T	Rockwell Superficial Hardness 45 kgf 1/16" ball HR45T
100	3.79	256	61.5	91.5	80.4	70.2
99	3.85	248	60.9	91.2	79.7	69.2
98	3.91	240	60.3	90.8	79.0	68.2
97	3.96	233	59.7	90.4	78.3	67.2
96	4.02	226	59.1	90.1	77.7	66.1
95	4.08	219	58.5	89.7	77.0	65.1
94	4.14	213	58.0	89.3	76.3	64.1
93	4.20	207	57.4	88.9	75.6	63.1
92	4.24	202	56.8	88.6	74.9	62.1
91	4.30	197	56.2	88.2	74.2	61.1
90	4.35	192	55.6	87.8	73.5	60.1
89	4.40	187	55.0	87.5	72.8	59.0
88	4.45	183	54.5	87.1	72.1	58.0
87	4.51	178	53.9	86.7	71.4	57.0
86	4.55	174	53.3	86.4	70.7	56.0
85	4.60	170	52.7	86.0	70.0	55.0
84	4.65	167	52.1	85.6	69.3	54.0
83	4.70	163	51.5	85.2	68.6	52.9
82	4.74	160	50.9	84.9	67.9	51.9
81	4.79	156	50.4	84.5	67.2	50.9
80	4.84	153	49.8	84.1	66.5	49.9

a. All relative hardness values in this table are averages of tests on various metals whose different properties prevent establishment of exact mathematical conversions. These values are consistent with ASTM A 370-91 for austenitic steels. It is recommended that ASTM standards A 370, E 140, E 10, E 18, E 92, E 110 and E 384, involving hardness tests on metals, be reviewed prior to interpreting hardness conversion values.

APPROXIMATE HARDNESS CONVERSION NUMBERS FOR NICKEL & HIGH-NICKEL ALLOYS

Vickers[a] HV	Brinell[b] HB	Rockwell Hardness Number[c] HRA	HRB	HRC	HRD	HRE	HRF	HRG	HRK
513	479	75.5	---	50.0	63.0	---	---	---	---
481	450	74.5	---	48.0	61.5	---	---	---	---
452	425	73.5	---	46.0	60.0	---	---	---	---
427	403	72.5	---	44.0	58.5	---	---	---	---
404	382	71.5	---	42.0	57.0	---	---	---	---
382	363	70.5	---	40.0	55.5	---	---	---	---
362	346	69.5	---	38.0	54.0	---	---	---	---
344	329	68.5	---	36.0	52.5	---	---	---	---
326	313	67.5	---	34.0	50.5	---	---	---	---
309	298	66.5	106	32.0	49.5	---	116.5	94.0	---
285	275	64.5	104	28.5	46.5	---	115.5	91.0	---
266	258	63.0	102	25.5	44.5	---	114.5	87.5	---
248	241	61.5	100	22.5	42.0	---	113.0	84.5	---
234	228	60.5	98	20.0	40.0	---	112.0	81.5	---
220	215	59.0	96	17.0	38.0	---	111.0	78.5	100.0

APPROXIMATE HARDNESS CONVERSION NUMBERS FOR NICKEL & HIGH-NICKEL ALLOYS (Continued)									
Vickers[a]	Brinell[b]	Rockwell Hardness Number[c]							
HV	HB	HRA	HRB	HRC	HRD	HRE	HRF	HRG	HRK
209	204	57.5	94	14.5	36.0	---	110.0	75.5	98.0
198	194	56.5	92	12.0	34.0	---	108.5	72.0	96.5
188	184	55.0	90	9.0	32.0	108.5	107.5	69.0	94.5
179	176	53.5	88	6.5	30.0	107.0	106.5	65.5	93.0
171	168	52.5	86	4.0	28.0	106.0	105.0	62.5	91.0
164	161	51.5	84	2.0	26.5	104.5	104.0	59.5	89.0
157	155	50.0	82	---	24.5	103.0	103.0	56.5	87.5
151	149	49.0	80	---	22.5	102.0	101.5	53.0	85.5
145	144	47.5	78	---	21.0	100.5	100.5	50.0	83.5
140	139	46.5	76	---	19.0	99.5	99.5	47.0	82.0
135	134	45.5	74	---	17.5	98.0	98.5	43.5	80.0
130	129	44.0	72	---	16.0	97.0	97.0	40.5	78.0
126	125	43.0	70	---	14.5	95.5	96.0	37.5	76.5
122	121	42.0	68	---	13.0	94.5	95.0	34.5	74.5
119	118	41.0	66	---	11.5	93.0	93.5	31.0	72.5
115	114	40.0	64	---	10.0	91.5	92.5	---	71.0
112	111	39.0	62	---	8.0	90.5	91.5	---	69.0
108	108	---	60	---	---	89.0	90.0	---	67.5
106	106	---	58	---	---	88.0	89.0	---	65.5
103	103	---	56	---	---	86.5	88.0	---	63.5
100	100	---	54	---	---	85.5	87.0	---	62.0
98	98	---	52	---	---	84.0	85.5	---	60.0
95	95	---	50	---	---	83.0	84.5	---	58.0
93	93	---	48	---	---	81.5	83.5	---	56.5
91	91	---	46	---	---	80.5	82.0	---	54.5
89	89	---	44	---	---	79.0	81.0	---	52.5
87	87	---	42	---	---	78.0	80.0	---	51.0
85	85	---	40	---	---	76.5	79.0	---	49.0
83	83	---	38	---	---	75.0	77.5	---	47.0
81	81	---	36	---	---	74.0	76.5	---	45.5
79	79	---	34	---	---	72.5	75.5	---	43.5
78	78	---	32	---	---	71.5	74.0	---	42.0
77	77	---	30	---	---	70.0	73.0	---	40.0

a. Vickers Hardness Number, Vickers indenter, 1.5, 10, 30-kgf load.

b. Brinell Hardness Number, 10 mm ball, 3000 kgf load. Note that in Table 5 of ASTM Test Method E 10, the use of a 3000-kgf load is recommended (but not mandatory) for material in the hardness range 96 to 600 HV, and a 1500-kgf load is recommended (but not mandatory) for material in the hardness range 48 to 300 HV. These recommendations are designed to limit impression diameters to the range 2.50 to 6.0 mm. The Brinell hardness numbers in this conversion table are based on tests using a 3000-kgf load. When the 1500-kgf load is used for the softer nickel and high-nickel alloys, these conversion relationships do not apply.

c. A Scale - 60-kgf load, diamond penetrator; B Scale - 100-kgf load, 1/16 in. (1.588 mm) ball; C Scale - 150-kgf load, diamond penetrator; D Scale - 100-kgf, diamond penetrator; E Scale - 100-kgf load, 1/8 in. (3.175 mm) ball; F Scale - 60-kgf load, 1/16 in. (1.588 mm) ball; G Scale - 150-kgf load, 1/16 in. (1.588 mm) ball; K Scale - 150-kgf load, 1/8 in. (3.175 mm) ball.

APPROXIMATE HARDNESS CONVERSION NUMBERS FOR NICKEL & HIGH-NICKEL ALLOYS							
Vickers[a]	Brinell[b]	Rockwell Superficial Hardness[c]					
HV	HB	HR15-N	HR30-N	HR45-N	HR15-T	HR30-T	HR45-T
513	479	85.5	68.0	54.5	---	---	---
481	450	84.5	66.5	52.5	---	---	---
452	425	83.5	64.5	50.0	---	---	---
427	403	82.5	63.0	47.5	---	---	---
404	382	81.5	61.0	45.5	---	---	---
382	363	80.5	59.5	43.0	---	---	---
362	346	79.5	58.0	41.0	---	---	---
344	329	78.5	56.0	38.5	---	---	---
326	313	77.5	54.5	36.0	---	---	---
309	298	76.5	52.5	34.0	94.5	85.5	77.0
285	275	75.0	49.5	30.0	94.0	84.5	75.0
266	258	73.5	47.0	26.5	93.0	83.0	73.0
248	241	72.0	44.5	23.0	92.5	81.5	71.0
234	228	70.5	42.0	20.0	92.0	80.5	69.0
220	215	69.0	39.5	17.0	91.0	79.0	67.0
209	204	68.0	37.5	14.0	90.5	77.5	65.0
198	194	66.5	35.5	11.0	89.5	76.0	63.0
188	184	65.0	32.5	7.5	89.0	75.0	61.0
179	176	64.0	30.5	5.0	88.0	73.5	59.5
171	168	62.5	28.5	2.0	87.5	72.0	57.5
164	161	61.5	26.5	-0.5	87.0	70.5	55.5
157	155	---	---	---	86.0	69.5	53.5
151	149	---	---	---	85.5	68.0	51.5
145	144	---	---	---	84.5	66.5	49.5
140	139	---	---	---	84.0	65.5	47.5
135	134	---	---	---	83.0	64.0	45.5
130	129	---	---	---	82.5	62.5	43.5
126	125	---	---	---	82.0	61.0	41.5
122	121	---	---	---	81.0	60.0	39.5
119	118	---	---	---	80.5	58.5	37.5
115	114	---	---	---	79.5	57.0	35.5
112	111	---	---	---	79.0	56.0	33.5
108	108	---	---	---	78.5	54.5	31.5
106	106	---	---	---	77.5	53.0	29.5
103	103	---	---	---	77.0	51.5	27.5
100	100	---	---	---	76.0	50.5	25.5
98	98	---	---	---	75.5	49.0	23.5
95	95	---	---	---	74.5	47.5	21.5
93	93	---	---	---	74.0	46.5	19.5
91	91	---	---	---	73.5	45.0	17.0
89	89	---	---	---	72.5	43.5	14.5
87	87	---	---	---	72.0	42.0	12.5
85	85	---	---	---	71.0	41.0	10.0
83	83	---	---	---	70.5	39.5	7.5
81	81	---	---	---	70.0	38.0	5.5
79	79	---	---	---	69.0	36.5	3.0
78	78	---	---	---	68.5	35.5	1.0

APPROXIMATE HARDNESS CONVERSION NUMBERS FOR NICKEL & HIGH-NICKEL ALLOYS (Cont.)							
Vickers[a] HV	**Brinell[b] HB**	**Rockwell Superficial Hardness[c]**					
		HR15-N	**HR30-N**	**HR45-N**	**HR15-T**	**HR30-T**	**HR45-T**
77	77	---	---	---	67.5	34.0	-1.5

a. Vickers Hardness Number, Vickers indenter, 1.5, 10, 30-kgf load.

b. Brinell Hardness Number, 10 mm ball, 3000 kgf load. Note that in Table 5 of ASTM Test Method E 10, the use of a 3000-kgf load is recommended (but not mandatory) for material in the hardness range 96 to 600 HV, and a 1500-kgf load is recommended (but not mandatory) for material in the hardness range 48 to 300 HV. These recommendations are designed to limit impression diameters to the range 2.50 to 6.0 mm. The Brinell hardness numbers in this conversion table are based on tests using a 3000-kgf load. When the 1500-kgf load is used for the softer nickel and high-nickel alloys, these conversion relationships do not apply.

c. 15-N Scale - 15-kgf load, superficial diamond penetrator; 30-N Scale - 30-kgf load, superficial diamond penetrator; 45-N Scale - 45-kgf load, superficial diamond penetrator; 15-T Scale - 15-kgf load, 1/16 in. (1.588 mm) ball; 30-T Scale - 30-kgf load, 1/16 in. (1.588 mm) ball; 45-T Scale - 45-kgf load, 1/16 in. (1.588 mm) ball.

Appendix 3

INTERNATIONAL STANDARDS ORGANIZATIONS, TECHNICAL ASSOCIATIONS & SOCIETIES

International Standards Organizations

AENOR	Asoiación Española de Normalización y Cetificación (Spain) tel +34 1 310 48 51, fax +34 1 310 49 76
AFNOR	Association Française de Normalisation (France) tel +33 1 42 91 55 55, fax +33 1 42 91 56 56
API	American Petroleum Institute (USA) tel +202 682 8375, fax +202 962 4776
ANSI	American National Standards Institute (USA) tel +212 642 4900, fax +212 302 1286
ASTM	American Society for Testing and Materials tel +610 832 9585, fax +610 832 9555
BSI	British Standards Institution (England) tel +44 181 996 70 00, fax +44 181 996 70 01
CSA	Canadian Standards Association (Canada) tel +416 747 4044, fax +416 747 2475
CSCE	Canadian Society for Chemical Engineers (613) 526-4652
DIN	Deutches Institut für Normung e.V. (Germany) tel +49 30 26 01 2260, fax +49 30 2601 1231
DS	Dansk Standard (Denmark) tel +45 39 77 01 01, fax +45 39 77 02 02
ELOT	Hellenic Organization for Standardization (Greece) tel +30 1 201 50 25, fax +30 1 202 07 76
IBN/BIN	Institut Belge de Normalisation/Belgisch Instituut voor Normalisatie (Belgium) tel +32 2 738 01 11, fax +32 2 733 42 64
IPQ	Instituto Português da Qualidade (Potugal) tel +351 1 294 81 00, fax +351 1 294 81 01
ISO	International Organization for Standardization (Switzerland) tel +41 22 749 01 11, fax +41 22 733 34 30
ITM	Inspection du Travail et des Mines (Luxembourg) tel +352 478 61 54, fax +352 49 14 47
JSA	Japanese Standards Association (Japan) tel +03 3583 8074, fax + 033582 2390
NBBPVI	National Board of Boiler and Pressure Vessel Inspectors (USA) tel +614 888 8320, fax +614 847 1828
NNI	Nederlands NormalisatieiInstituut (Netherlands) tel +31 15 69 03 90, fax +31 15 69 01 90

International Standards Organizations (Continued)

NSAI	National Standards Authority of Ireland tel +353 1 837 01 01, fax +353 1 836 98 21
NFS	Norges Standardiseringsforbund (Norway) tel +47 22 46 60 94, fax +47 22 46 44 57
ON	Österreichisches Normungsindtitut (Austria) tel +43 1 213 00, fax +43 1 213 00 650
SA	Standards Australia tel +08 373 1540, fax +08 373 1051
SCC	Standards Council of Canada tel +800 267 8220, fax +613 995 4564
SIS	Standardiseringskommissione n i Sverige (Sweden) tel +46 8 613 52 00, fax +46 8 411 70 35
SFS	Suomen Standardisoimisliitto r.y. (Findland) tel +358 0 149 93 31, fax +358 0 146 49 25
SNV	Schweizerische Normen-Vereinigung (Switzerland) tel +41 1 254 54 54, fax +41 1 254 54 74
STRI	Technological Institute of Iceland tel +354 587 70 02, fax +354 587 74 09
UNI	Ente Nazionale Italiano di Unificazione (Italy) tel +39 2 70 02 41, fax +39 2 70 10 61 06

Technical Associations, Societies, and Institutes[a]

AA	The Aluminum Association (202) 862-5100
AEE	The Association of Energy Engineers (404) 447-5083
AFS	American Foundrymen's Society (312) 824-0181
AISI	Association of Iron and Steel Engineers (412) 281-6323
AlChE	American Institute of Chemical Engineers (212) 705-7338
AMEC	Advanced Materials Engineering Centre (902) 425-4500
API	American Petroleum Institute (202) 682 8375
ASEE	American Society for Engineering Education (202) 331-3500
ASM	ASM International - The Materials Information Society (800) 336-5152 or (216) 338-5151
ASME	American Society of Mechanical Engineers (212) 705-7722
ASNT	American Society for Nondestructive Testing (614) 274-6003
ASQC	American Society for Quality Control (414) 272-8575
ASTM	American Society for Testing and Materials (610) 832-9585
AWI	American Welding Institute (615) 675-2150
AWS	American Welding Society (305) 443-9353 or (800) 443-9353
CAIMF	Canadian Advanced Industrial Materials Forum (416) 798-8055
CASI	Canadian Aeronautic & Space Institute (613) 234-0191
CCA	Canadian Construction Association (613) 236-9455
CCPE	Canadian Council of Professional Engineers (613) 232-2474
CCS	Canadian Ceramics Society (416) 491-2886
CDA	Copper Development Association (212) 251-7200
CEN	European Committee for Standardization +32 2 550 08 11
CIE	Canadian Institute of Energy (403) 262-6969
CIM	Canadian Institute for Mining and Metallurgy (514) 939-2710
CMA	Canadian Manufacturing Association (416) 363-7261

Technical Associations, Societies & Institutes[a] (Continued)

CNS	Canadian Nuclear Society (416) 977-6152
CPI	Canadian Plastics Institute (416) 441-3222
CPIC	Canadian Professional Information Centre (905) 624-1058
CSEE	Canadian Society of Electronic Engineers (514) 651-6710
CSME	Canadian Society of Mechanical Engineers (514) 842-8121
CSNDT	Canadian Society for Nondestructive Testing (416) 676-0785
El	Engineering Information Inc. (212) 705-7600
FED	Federal & Military Standards (215) 697-2000
IEEE	lnstitute of Electrical & Electronic Engineers (212) 705-7900
IES	Institute of Environmental Sciences (312) 255-1561
IIE	Institute of Industrial Engineers (404) 449-0460
IMMS	International Material Management Society (705) 525-4667
ISA	Instrument Society of America (919) 549-8411
ISA	Instrument Society of America (919) 549-8411
ISS	Iron and Steel Society (412) 776-1535
ITI	International Technology Institute (412) 795-5300
ITRI	International Tin Research Institute (614) 424-6200
MSS	Manufactures Standardization Society of Valves & Fittings Industry (703) 281-6613
MTS	Marine Technology Society (202) 775-5966
NACE	National Association of Corrosion Engineers (713) 492-0535
NAPE	National Association of Power Engineers (212) 298-0600
NAPEGG	Association of Professional Engineers, Geologists and Geophysicists of theNorthwest Territories (403) 920-4055
NBBPVI	National Board of Boiler and Pressure Vessel Inspectors 614 888 8320
NiDI	Nickel Development Institute (416) 591-7999
PIA	Plastics Institute of America (201) 420-5553
RIA	Robotic Industries Association (313) 994-6088
SAE	Society of Automotive Engineers (412) 776-4841
SAME	Society of American Military Engineers (703) 549-3800
SAMPE	Society for the Advancement of Materials and Processing Engineering (818) 331-0616
SCC	Standards Council of Canada (800) 267-8220
SCTE	Society of Carbide & Tool Engineers (216) 338 5151
SDCE	Society of Die Casting Engineers (312) 452-0700
SME	Society of Manufacturing Engineers (313) 271-1500
SPE	Society of Petroleum Engineers (214) 669-3377
SSIUS	Specialty Steel Industry of the United States (202) 342-8630
SSPC	Steel Structures Painting Council (412) 268-3327
STC	Society for Technical Communications (202) 737-0035
STLE	Society of Tribologists and Lubrication Engineers (312) 825-5536
TDA	Titanium Development Association (303) 443-7515
TMS	The Minerals, Metals, and Materials Society (412) 776-9000
WIC	Welding Institute of Canada (905) 257-9881
WRC	Welding Research Council (212) 705-7956

a. Telephone numbers only.

SUBJECT INDEX

B

C

D

E

M

P

R

S

T

U

V

W

CODE PARAGRAPH INDEX